OCR Science for GCSE

Separate Award

Biology

Byron Dawson

Ian Honeysett

Paul Spencer

Series editor: Bob McDuell

www.heinemann.co.uk

✓ Free online support
✓ Useful weblinks
✓ 24 hour online ordering

01865 888058

Heinemann

Inspiring generations

Heinemann is an imprint of Pearson Education Limited, a company incorporated in England and Wales, having its registered office at Edinburgh Gate, Harlow, Essex, CM20 2JE. Registered company number: 872828

Heinemann is a registered trademark of Pearson Education Limited

First published 2006
© Harcourt Education Limited, 2006

10 09
10 9 8 7 6 5 4

978 0 435 67526 4

Series Editor: Bob McDuell
Designed by Wooden Ark
Typeset by HL Studios, Long Hanborough, Oxford

Original illustrations © Harcourt Education Limited, 2006

Illustrated by HL Studios
Cover design by Cooney Bains
Printed in China (CTPS/04)
Cover photo: © Getty Images
Picture research by Kay Altwegg and Ginny Stroud-Lewis

Acknowledgements
The authors and publisher would like to thank the following individuals and organisations for permission to reproduce photographs:

Page 2, Getty Images; 3, Ian Hooton / SPL; 5, Michael Donne / SPL (x3); 5, BR Aaron Haupt / SPL; 6, Alamy Images / Sally and Richard Greenhill; 7, Getty Images / Food Pix; 8, Andy Crump, TDR, WHO / SPL; 10, Corbis / Brooke Fasani; 11, Alamy Images / David Hoffman Photo Library; 12, T Martin Dohrn / SPL; 12, B Corbis / Sygma / Rogate Joe; 13, SPL / BSIP, LA; 14, Corbis / Greg Smith; 15, Getty Images / Photodisc; 18, Author supplied; 18, B Rory McClenaghan / SPL; 19, T Pearson Education / Ginny Stroud-Lewis; 19, B SPL / Tek Image; 20, Wellcome Medical Photographic Library; 22, Alamy Images / Alex Segre; 23, T Karsten Wrobel / Alamy; 23, B Brian Snyder / Reuters / Corbis; 26, Corbis; 27, Getty Images / PhotoDisc; 28, Author supplied; 29, SPL / Dept. of Clinical Cytogenetics, Addenbrookes Hospital; 30, www.genome.gov; 31, T Alamy Images / Janine Wiedel Photolibrary; 31, B Dr Jeremy Burgess / SPL; 33, T Alamy Images; 33, B Dept. Of Clinical Cytogenetics, Addenbrookes Hospital / SPL; 34, SPL / Simon Fraser / RVI, Newcastle-Upon-Tyne; 38, The Kobal Collection / Amblin / Universal; 39, T Image Quest; 39, B Alamy Images / Glyn Thomas; 41, T Education Photos / Alamy; 41, B UK Ladybird Survey / Peter Brown; 42, Alamy; 43, T Museum Images / Pierre Fidenci; 43, B Getty Images / Photodisc; 44, T SPL; 44, M Alamy; 44, B Getty Images / Photodisc; 45, T Getty Images / Photodisc; 45, B Corbis; 46, Alamy / Royalty Free; 47, OSF; 48, T Paola Zucchi / ABPL; 48, B Getty Images / Photodisc; 50, Getty Images / PhotoDisc; 51, T Alamy Images / Motoring Picture Library; 51, B Corbis / Juan Medina / Reuters; 52, T The Photolibrary Wales / Alamy; 52, B Corbis; 53, TR K.H. Kjeldsen / SPL; 53, TL Mediscan; 53, B ImageState / Alamy; 54, Alamy Images / Andrew Darrington; 55, T Getty Images / Photodisc; 55, B Corbis / Chris Mattison; Frank Lane Picture Agency; 56, T Corbis; 56, B NHPA / Mike Lane; 57, Getty Images / Photodisc; 58, SPL / B. Murton / Southampton Oceanography Centre; 59, T Getty Images / Botanica / Emily Brooke Sandor; 59, B SPL / Sinclair Stammers; 60, T Corbis / Bettmann; 60, B OSF / Peter Parks; 61, T OSF / Peter Parks; 61, B OSF / David Fox; 62 Getty Images / Hulton

Archive; 63, T Alamy Images / PCL; 63, B Alamy Images / David R. Frazier Photolibrary, Inc.; 64 NASA / SPL; 65, T Michael Donne / SPL; 65, ML Adrian Davies / naturepl.com; 65, MR Ken Preston-Mafham / premaphotos.com; 65, B Dr Jeremy Burgess / SPL; 66, T SPL / Dr Jeremy Burgess; 66, B Natural Visions / Heather Angel; 67, T Doug Perrine / naturepl. com; 67, M Roger Key / English Nature; 67, B Mark Carwardine / naturepl.com; 68, T Paul Glendell / Alamy; 68, B NHPA; 69, NHPA / Trevor McDonald; 70, T Corbis / Amos Nachoum; 70, B f1 online / Alamy; 74, Corbis; 75, T SPL / J.C. Revy; 75, M Eric Grave / Phototake Inc / OSF; 75, B SPL / A. Barrington Brown; 78, Getty Images / Photodisc; 79, T NHPA / Andrea Bonetti; 79, B Getty Images / Photodisc; 80, SPL / Steve Gschmeissner; 82, Getty Images / Photodisc; 83, Pearson Index; 85, The Advertising Archives; 86, T Getty Images / Photodisc; 86, B Corbis / Bettmann; 87, T SPL / Du Cane Medical Imaging Ltd; 87, B Pascal Goetgheluck / SPL; 89, Getty Images / Photodisc; 90, Getty Images / Photodisc; 91, Corbis; 94, SPL / Professor Miodrag Stojkovic; 95, T Corbis / Frank Young / Papilio; 95, M & B www.mykoweb.com; 97, Corbis / Owen Franken; 99, T Stone / Getty; 99, B: SPL / Cristina Pedrazzini; 100, Chris Rogers / Corbis; 101, Corbis / James Darling / Reuters; 102, Art Directors and Trip; 103, T Roslin Institute / Phototake Inc / OSF; 103, B Eyewire; 106, Getty Images / Photodisc; 110, Getty Images / Photodisc; 111, T Corbis; 111, M Pearson Index; 111, B SPL / Dr Jeremy Burgess; 113, Richard Smith; 114, Art Directors and Trip; 115, Holt Studios International Ltd / Alamy (x2); 116, L Alamy Images / Phototake Inc; 116, R Steve Gschmeissner / SPL; 118, Getty Images / Photodisc (x2); 119, T SPL / Volker Steger; 119, B Art Directors and Trip; 121, L SPL / J.C. Revy; 121, R SPL / Andrew Syred; 122, SPL / Martyn F. Chillmaid; 123, T Pearson Index; 123, B Mary Ellen Baker / Botanica / OSF; 124, cumulus; 125, Alamy; 126, SPL / Dr Jeremy Burgess; 127, Pearson Education / Ginny Stroud-Lewis (x2); 130, Alamy Images / Andre Jenny; 131, T Stone / Getty; 131, B Alamy Images / FLPA; 133, T Mary Clark / Alamy; 133, B Harcourt Education / Ginny Stroud-Lewis; 134, T Pearson Index; 134, B Corbis / Joe McDonald; 135, T Corbis / Corbis Sygma; 135, B Pearson Education / Ginny Stroud-Lewis; 136, T Hybrid Medical Animation / SPL; 136, B Robert Pickett / Corbis; 137, Pearson Education / Ginny Stroud-Lewis; 138, Alamy Images / Hugh Threlfall; 138; SPL; 140, SPL / Andrew McClenaghan; 141, Alamy Images / Elmtree Images; 142, T Dr Jeremy Burgess / SPL; B SPL / Michael Abbey; 146, Pearson Education; 147, T Pearson Education; 147, BL Anatomical Travelogue / SPL; 147, BR Maximilian Stock Ltd / SPL; 149, James Stevenson / SPL; 150, TL Sovereign, ISM / SPL; 150, BL BSIP / SPL; 150, M Pearson Index; 150, R James King-Holmes / SPL; 151, Antonia Reeve / SPL; 153, Scott Camazine / SPL; 154, Gusto / SPL; 155, Popperfoto / Alamy; 156, T Athenais, ISM / SPL; 156, B SPL; 157, T D. Phillips / SPL; 157, M1 Steve Gschmeissner / SPL; 157, M2 Biology Media / SPL; 157, B NIBSC / SPL; 158, Pearson Index, 159, Alfred Pasieka / SPL; 160, BSIP, CIOT / SPL; 163, Nik Milner / Syner-Comm / Alamy; 166, Nik Milner / Syner-Comm / Alamy; 167, Pearson Index; 169, Bettmann / Corbis; 170, Mauro Fermariello / SPL; 171, Princess Margaret Rose Orthopaedic Hospital / SPL; 172, T Neil Tingle / action plus; 172, B Antonia Reeve / SPL; 173, NHS UK Transplant; 174, BananaStock / Alamy; 175, T Pearson Index, 175, B Michael Pitts / naturepl.com; 176, Bettmann / Corbis; 177, Earl & Nazima Kowall / Corbis; 178, Peter Beck / Corbis; 182, TEK Image / SPL; 183, T Food Features; 183, B SCIMAT / SPL; 184, Dr Linda Stannard, UCT / SPL; 185, Bob Krist / Corbis; 186, Pearson Education Ltd. / Trevor Clifford; 187, TR David Young-Wolff / Alamy; 187, BR Custom Medical Stock Photo / SPL; 187, ML SPL; 187, MR St Mary's Hospital Medical School / SPL; 188, L Eye Of Science / SPL (x2); 188, M Daniel Leclair / Reuters / Corbis; 189, Scott Camazine / Alamy; 190, Jorgen Schytte / Still Pictures; 191, T Pearson Index; 191, B SCIMAT / SPL; 192, Anuruddha Lokuhapuarachchi / Reuters / Corbis; 193, Macduff Everton / Corbis; 194, Royalty-Free / Corbis; 195, Pearson Education; 196, geogphotos / Alamy; 198, L Ricardo Funari / BrazilPhotos / Alamy; 198, R Eye Ubiquitous / Alamy; 199, Dietmar Nill / naturepl.com; 200, T Alamy - Stacey Richards; 200, M Russ Munn / AGSTOCKUSA / SPL; 200, B Nigel Cattlin / FLPA; 202, T Kim Taylor / naturepl.com; 202, B Stephen Dalton / NHPA; 203, T Network Photographers / Alamy; 203, BL Fabio Liverani / naturepl.com; 203, BR Kim Taylor / naturepl.com; 204, Kim Taylor / naturepl.com; 206, Edward Parker / Alamy; 207, T Image Source / Alamy; 207, B Cordelia Molloy / SPL; 208, Iconotec / Alamy; 209, Daniel Heuclin / NHPA; 210, Sian Irvine / Anthony Blake Photo Library; 211, Associated Press; 212, Julia Kamlish / SPL; 213, Holt Studios International Ltd / Alamy; 214, T Planetary Visions Ltd / SPL; 214, B Norm Thomas / SPL; 219, Corbis; 219, Getty Images/Photodisc (x2)

The authors and publisher would like to thank the following individuals and organisations for permission to reproduce copyright materials:

49 T (photosynthesis graph) and 72 L (Predator/prey relationship), www.bbc.co.uk/gcsebitesize; 93, TL Relative rates graph for the human body,- Advanced Biology Principles and Applications by Clegg and MacKean, published by John Murray / BR Growth curves for boys head size and weight - Child Growth Foundation; 154 B (ECG traces for two people) Harper Collins (pub 1995) Collins Vocational Sciences, Biology Pack for Advanced GNVQ - Article by S. Spencer; 162 T/B (onset of asthma diagram) Collins Vocational Sciences, Biology Pack for Advanced GNVQ ISBN 0 00 322376 0 -article by P Spencer (x2); 173 T (NHS Organ Donor Register) NHS; 174 M (Lung transplant graph/patient survial data) www.mayoclinic.org; 176 B (girls height curve/combined girls and boys height) Human Biology Collins Advanced Science ISBN 0 00 3290956 - Boyle, Indge and Senior (x2); 205 T (The Antartic Food web) Collins Advanced Science, Biology ISBN 000 322 3272.

Every effort has been made to contact copyright holders of material reproduced in this book. Any omissions will be rectified in subsequent printings if notice is given to the publishers.
Tel: 01865 888058   www.heinemann.co.uk

Introduction

This student book covers the new OCR Gateway Biology specification for Higher tier. It has been written to support you as you study for the OCR Gateway Biology GCSE.

This book has been written by examiners who are also teachers and who have been involved in the development of the new specification. It is supported by other material produced by Heinemann, including online teacher resource sheets and interactive learning software with exciting video clips, games and activities.

As part of GCSE Gateway Biology you must complete either:

a Can-do tasks in biology
Science in the news

OR

b Research study
Data task
Teacher assessment of your practical skills.

These are fully explained on pages 220–229.

We hope this book will help you achieve the best you can in your GCSE Biology Award and help you understand how much biology affects our everyday lives. As citizens of the 21st century you need to be informed about biological issues. Then you can read newspapers or watch television programmes and really have views about things that affect you and your family.

Gateway Biology GCSE is a very good preparation for AS courses in Biology.

The next two pages explain the special features we have included in this book to help you to learn and understand the subject, and to be able to use it in context. At the back of the book you will also find some useful tables, as well as a glossary and index.

About this book

This student book has been designed to make learning biology fun. The book follows the layout of the OCR Gateway specification. It is divided into six sections that match the six modules in the specification for Biology: B1, B2, B3, B4, B5 and B6.

The module introduction page at the start of a module introduces what you are going to learn. It has some short introductory paragraphs, plus 'talking heads' with speech bubbles that raise questions about what is going to be covered.

Each module is then broken down into eight separate items (a–h), for example, B1a, B1b, B1c, B1d, B1e, B1f, B1g, B1h.

Each 'item' is covered in four book pages. These four pages are split into three pages covering the science content relevant to the item plus a 'context' page which places the science content just covered into context, either by news-related articles or data tasks, or by examples of scientists at work, science in everyday life or science in the news.

Throughout these four pages there are clear explanations with diagrams and photos to illustrate the biology being discussed. At the end of each module there are three pages of questions to test your knowledge and understanding of the module.

There are three pages of exam-style end of module questions for each module.

The talking heads on the module introduction page raise questions about what you are going to learn.

The numbers in square brackets give the marks for the question or part of the question.

The bulleted text introduces the module.

This box highlights what you need to know before you start the module.

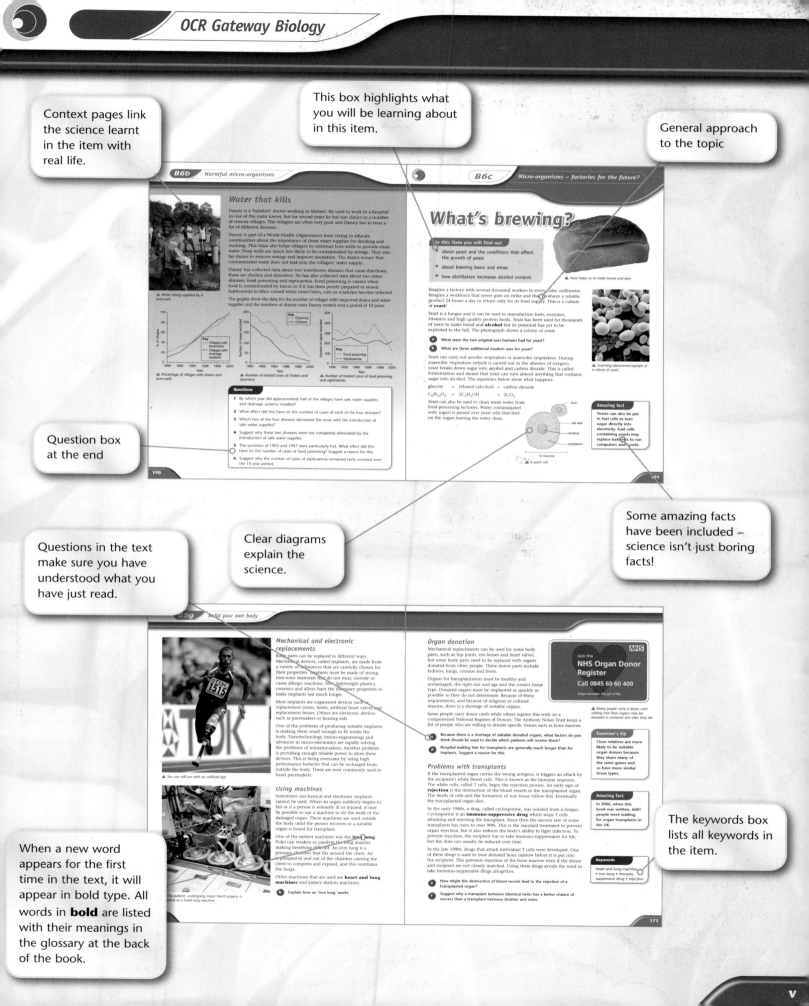

Context pages link the science learnt in the item with real life.

This box highlights what you will be learning about in this item.

General approach to the topic

Question box at the end

Questions in the text make sure you have understood what you have just read.

Clear diagrams explain the science.

Some amazing facts have been included – science isn't just boring facts!

When a new word appears for the first time in the text, it will appear in bold type. All words in **bold** are listed with their meanings in the glossary at the back of the book.

The keywords box lists all keywords in the item.

Contents

B1 Understanding ourselves

Being fit can be enjoyable. You feel so much better when you are healthy and fit.

I can't be bothered to get fit. I am healthy and you have got to die of something so I may as well enjoy life.

- This unit is about understanding ourselves. It is only by understanding who and what we are that we can learn to make the right decisions that will keep us healthy and help us to enjoy a long and happy life. In this module you will learn how we use food to obtain energy and what happens in parts of the world where food is scarce and people are starving. You will also learn how we are affected by disease and what we can do to keep healthy.

- One of the reasons that humans have been so successful and are found living all over the world is that our bodies can regulate and keep constant many of our internal processes, such as body temperature. However, these internal mechanisms can be altered by drugs. Some people's lives have been ruined by drugs. You will learn about different types of drugs and the affect they can have on the body.

- Finally, you will learn about DNA and how it makes us who and what we are. The study of DNA is an exciting new branch of science and it promises to bring about many new and wonderful changes in your lifetime.

What you need to know

- About types of food, what it contains and that it provides us with energy.

- How we can stay fit and healthy.

- Microbes can cause disease.

- Variation exists in animals and plants and that we are all different.

Fighting fit

In this item you will find out

- about your blood pressure and what happens if it gets too high or too low

- about the difference between fitness and health and how fitness can be measured

- about aerobic and anaerobic respiration

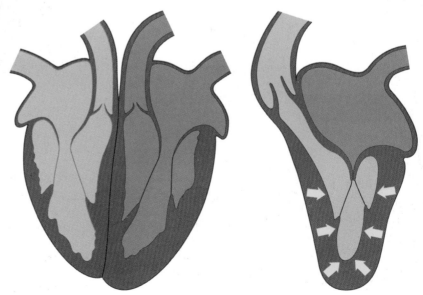

Have you ever had your blood pressure taken by a doctor? When the doctor gives you the result you get two readings instead of one.

Your heart beats about 80 times every minute. When it beats, blood is forced through the blood vessels and your blood pressure is at its highest. This is called the **systolic** pressure and is the first number given.

When your heart is resting the narrow blood vessels slow the blood flowing through them, which lowers the pressure. This is called the **diastolic** pressure and is the second number given. Blood pressure is measured in units called mm Hg.

Your blood pressure changes all of the time. It depends on how active you are, your age, weight and lifestyle, how much alcohol you drink and whether you are angry or calm.

▶ 'Your blood pressure is fine, 130 over 75'

 a Explain the difference between systolic and diastolic pressure.

Amazing fact

A young fit person may have a blood pressure of about 120 mm Hg over 70 mm Hg.

▲ Heart resting　　　　　　　　　　▲ Heart contracting

High and low pressure

If your blood pressure is too high or too low this can cause problems. High blood pressure can cause weak blood vessels to burst. If this happens in the brain, it can cause a 'stroke' which can damage your brain.

High blood pressure can also damage organs such as the kidneys. Some people have low blood pressure. This can lead to poor blood circulation, dizziness and fainting as the brain does not get enough oxygenated blood. The table shows normal blood pressure and high and low blood pressure.

Blood pressure	Systolic (mm Hg)	Diastolic (mm Hg)
normal	130	75
high	160	105
low	90	40

Aerobic respiration

Your body cells get energy by reacting glucose with oxygen. This is called **aerobic respiration**.

▶ Aerobic respiration

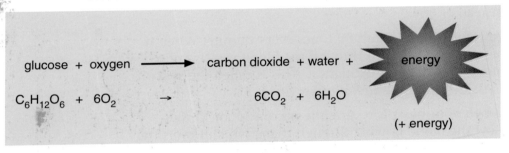

glucose + oxygen ⟶ carbon dioxide + water + energy

$$C_6H_{12}O_6 + 6O_2 \rightarrow 6CO_2 + 6H_2O$$

(+ energy)

Anaerobic respiration

When you do vigorous exercise your heart and lungs cannot provide your muscles with enough oxygen quickly enough. When this happens, your cells carry out **anaerobic respiration** as well as aerobic respiration.

▶ Anaerobic respiration

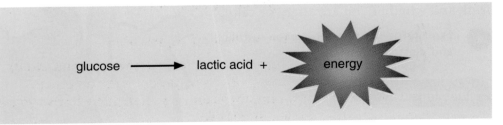

glucose ⟶ lactic acid + energy

Examiner's tip

The equation for respiration is the same as the equation for photosynthesis in reverse.

As well as energy, **lactic acid** is also produced during anaerobic respiration due to the incomplete breakdown of glucose. This lactic acid can build up in your muscles and cause muscle fatigue and pain. Anaerobic respiration does not produce as much energy as aerobic respiration.

b Describe three differences between aerobic and anaerobic respiration.

c Suggest why our bodies do not use anaerobic respiration all of the time.

Recovering from fatigue

When you do hard exercise it induces a lack of oxygen in your cells. This is called the **oxygen debt** and it has to be repaid.

When you respire anaerobically, glucose is broken down into lactic acid instead of carbon dioxide and water. When you stop exercising you are usually panting. You continue to breathe heavily until your lungs have provided your body with enough oxygen to break down all of the lactic acid into carbon dioxide and water. The carbon dioxide is then carried in the blood to the lungs where you breathe it out.

Your heart rate is also increased and this helps your blood to carry lactic acid to your liver where it can be broken down.

▲ Breathing and heart rate increase to deliver glucose and oxygen to the muscles

Fitness

Fitness is not the same as being healthy and free from disease. Fitness is how efficiently your body can perform some of its functions. It is usually a result of exercising.

▲ Lactic acid builds up causing fatigue and pain

▲ This person is well but not fit

▲ This person is fit but not well

d The overweight man is well but not healthy. Suggest why.

There are different ways of measuring fitness. You can measure how strong a person is by seeing how many press-ups they can do, and you can measure stamina by seeing how long they can keep doing an exercise. You can also measure agility, flexibility and speed.

Keywords

aerobic respiration •
anaerobic respiration •
diastolic pressure • lactic
acid • oxygen debt •
systolic pressure

The hidden problem

Nearly one in three people has high blood pressure. About 30% of people with high blood pressure are not aware that they have it. It tends to be older people who have higher blood pressure.

High blood pressure is defined as being higher than 140 mm Hg over 90 mm Hg. If high blood pressure is diagnosed it can be successfully treated in most people. If it is left untreated it can cause a variety of problems.

The graph shows the incidence of high blood pressure in different groups of people.

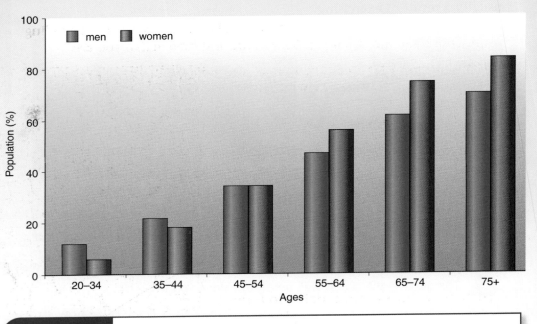

Questions

1 What proportion of the population has high blood pressure?

2 Does a person with a blood pressure of 140 mm Hg over 85 mm Hg have high blood pressure? Explain your answer.

3 Why is it important to diagnose high blood pressure early?

4 Which part of the population tends to have the highest blood pressure?

5 Which group of the population has shown the least change in the incidence of high blood pressure as they get old?

6 Average blood pressure fell between 1976 and 1994. Suggest why.

Diet dilemmas

In this item you will find out

- what is meant by a balanced diet and about protein intake
- about the digestive system
- how to calculate whether you are over or underweight

ONLY A YEAR TO EAT ALL THIS!

Amazing fact

Each year you eat about 500 kg of food. That's about the weight of 20 sacks of potatoes.

Do you eat a healthy balanced diet with lots of fruit and vegetables, meat, fish and whole grains?

Eating a balanced diet can be complicated because it can vary depending on how old you are, what gender you are and how active you are.

- young people who are growing need to eat more than older people who have stopped growing
- active people need to eat more than people who are not active
- men need to eat more than women.

There are also other factors that influence the types of food that people eat.

Vegetarians and vegans choose not to eat meat so they need a different kind of balanced diet from people who do eat meat.

People sometimes avoid certain foods for religious reasons and some people can have medical reasons for avoiding certain types of food, for example people who are allergic to peanuts.

a Why do you think that men need to eat more than women?

What's in our food?

Example of type of food	Food group	What we digest it into
Eggs	Protein	Amino acids
Butter	Fat	Fatty acid and glycerol
Potato	Carbohydrate	Simple sugars such as glucose

Animal proteins are called 'first class proteins' because they contain all the essential amino acids we need, but which we can't make for ourselves.

Most plant proteins only contain some of the amino acids that we need. Different plant proteins contain different amino acids.

 Suggest how a vegetarian could get all of the amino acids that they need.

We can calculate our recommended average daily protein intake by using the following formula:

RDA in g = 0.75 × body mass in kg

(RDA = recommended daily average)

 Emma weighs 48 kg. How much protein should she eat each day?

If we do not get enough protein in our diet, then we could suffer from protein deficiency (**kwashiorkor**). People in developing countries often suffer from it.

▶ *This child is suffering from a lack of protein in his diet*

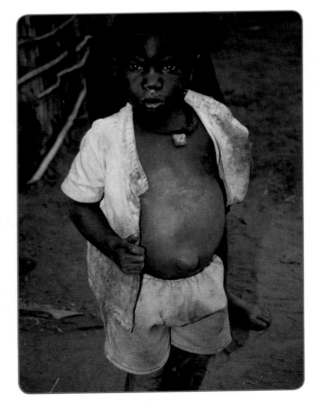

Digestion

When food is eaten, muscles in the gut contract and push the food along. As the food is pushed from the mouth to the anus, it is broken down by digestive **enzymes** into much smaller molecules. These small molecules can then be absorbed through the gut wall and into the blood plasma or the lymph to be used by the body. This is called **chemical digestion**.

 Suggest why food can only be absorbed when it is broken down into small molecules.

Food group	Digestive enzyme
Protein	**Proteases**
Fat	**Lipases**
Carbohydrate	**Carbohydrases**

Different food groups are digested by different enzymes in our mouths, stomachs and small intestines.

Food is kept in the stomach for several hours. During this time, hydrochloric acid is added to the food. This kills most of the bacteria on the food and helps to break it up into smaller molecules. Acid in the stomach also helps the enzymes to work.

e **Why is it important to destroy most of the bacteria found on our food?**

Food then passes into the small intestine where more enzymes are added to the food.

Bile is also added to the food in the small intestine. This improves fat digestion. Bile is not an enzyme. It is a chemical produced by the liver and stored in the gall bladder. When it is added to food, it makes the fat break up into smaller droplets. It works in the same way as washing up liquid breaks up fat when you wash greasy plates. This is called **emulsification**. This increases the surface area of the fat for the enzymes to work on.

f **Suggest why breaking fat up into smaller droplets makes it easier for enzymes to break down the fat.**

The small molecules of food are then absorbed through the walls of the small intestine and into the blood plasma or the lymph. This process happens by **diffusion**.

mouth
salivary gland
food
liver
bile duct
gall bladder
stomach
pancreas
small intestine
large intestine
anus

▲ The human digestive system

Keywords

bile • carbohydrase • chemical digestion • diffusion • emulsification • enzyme • kwashiorkor • lipase • protease

Body image

Sarah is 15. She spends a lot of time reading fashion magazines and she is very depressed about how she looks. She has been teased by some of the girls at her school. She wishes that she could look like one of the slim and attractive models she sees in the magazines then perhaps she wouldn't be bullied at school. She has started to eat less food. At first her parents didn't notice but she has lost a lot of weight and they are becoming worried.

They take her to see the doctor. Dr Mackay examines Sarah and explains to her that depending on sex, age and height, we all have an ideal weight. Dr Mackay tells Sarah that dieting can lead to problems such as anorexia, and being very underweight can have severe health risks. She suggests that Sarah talks to a counsellor about her feelings.

Sarah can work out her ideal weight by using the following formula to calculate her body mass index (BMI).

$$BMI = \text{mass in kg}/(\text{height in m})^2$$

She can then compare her BMI with the table to see if she is at a normal weight or is underweight or overweight.

Mass (kg)

Height (cm)	54	59	64	68	73	77	82	86	91	95	100	104	109	113
137	29	31	34	36	39	41	43	46	48	51	53	56	58	60
142	27	29	31	34	36	38	40	43	45	47	49	52	54	56
147	25	27	29	31	34	36	38	40	42	44	46	58	50	52
152	23	25	27	29	31	33	35	37	39	41	43	45	47	49
158	23	24	26	27	29	31	33	35	37	38	40	42	44	46
163	21	22	24	26	28	29	31	33	34	36	38	40	41	43
168	19	21	23	24	26	27	29	31	32	34	36	37	39	40
173	18	20	21	23	24	26	27	29	30	32	34	35	37	38
178	17	19	20	22	23	24	26	27	29	30	32	33	35	36
183	16	18	19	20	22	23	24	26	27	28	30	31	33	34
188	16	17	18	19	21	22	23	24	26	27	28	30	31	32
193	15	16	17	18	20	21	22	23	24	26	27	38	29	30
198	14	15	16	17	19	20	21	22	23	24	25	27	28	29
203	13	14	15	17	18	19	20	21	22	23	24	25	26	28

underweight   healthy weight   overweight   obese

Questions

1 Why does Sarah want to be thin?

2 Sarah weighs 54 kg and is 1.73 m tall. Calculate her BMI.

3 Look at the table. Is Sarah underweight, overweight or normal weight?

4 What may happen if Sarah loses too much weight?

5 Why did Dr Mackay suggest that Sarah sees a counsellor?

Fighting disease

In this item you will find out

- about different things that can make us ill

- how our body responds to illness

- how drugs are tested to make sure they are effective and safe

Illness can be caused by many different things.

Cancer occurs when body cells continue to divide uncontrollably. They produce a mass of cells called a **tumour**. Some tumours stop growing and are called **benign**. It is the tumours that continue to grow and spread that are dangerous. These are called **malignant**.

We can make changes to our lifestyles which may reduce the risk of getting some cancers. We can avoid too much sunshine, which contains UV light, smoking or certain types of food. If we eat lots of fresh fruit and vegetables containing antioxidants, this may reduce our chances of getting cancer.

The table shows the estimated survival rates after five years for men diagnosed with different types of cancer.

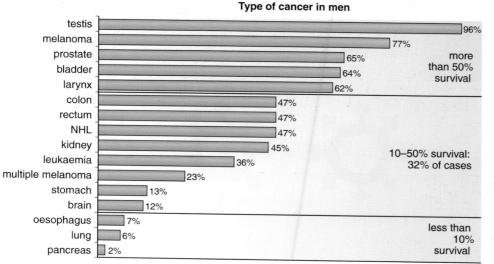

▲ *Causes of disease*

Type of cancer in men

Type	Survival
testis	96%
melanoma	77%
prostate	65%
bladder	64%
larynx	62%
colon	47%
rectum	47%
NHL	47%
kidney	45%
leukaemia	36%
multiple melanoma	23%
stomach	13%
brain	12%
oesophagus	7%
lung	6%
pancreas	2%

more than 50% survival

10–50% survival: 32% of cases

less than 10% survival

a Suggest which type of cancer is most easily treated in men.

b Suggest which type of cancer is the most difficult to treat in men.

Malaria

Malaria is one of the world's biggest killers. It is caused by a **parasite** that lives in the blood and liver. The human body and other animals act as a **host** to this parasite. The parasite is spread by a mosquito.

The mosquito sucks up the blood of an animal that has the parasite. The mosquito then carries the parasite and when it sucks the blood of a human, the parasite passes into the person's blood stream and causes malaria. Animals, such as the mosquito, that carry disease-causing organisms from one animal to another, are called **vectors**.

▲ *Mosquitoes spread malaria*

Controlling vectors

Diseases, such as malaria, can be prevented by destroying the mosquito vectors. Mosquitoes breed in water such as ponds and even puddles.

They can be killed by spraying them with chemicals, or by putting fish into the ponds that eat the mosquito larvae. People can take drugs to kill the parasite in their bodies, or use nets over their beds so that the mosquitoes cannot bite them during the night.

c **Suggest how other vectors such as the housefly could be controlled.**

Antibodies and antigens

Microorganisms cause disease when they damage the cells in our bodies or produce poisonous chemicals called **toxins**. Microorganisms that do this are called **pathogens** and each pathogen has its own **antigens**.

Our bodies respond by producing different **antibodies** for each antigen. The antibody locks onto the antigen and kills the pathogen.

▲ *Mosquitoes can be killed by chemicals*

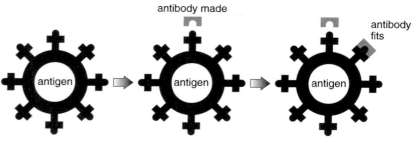

antibody made

antibody fits

antigen → antigen → antigen

▲ *How antibodies attack antigens*

Immunity

The problem is that it can take several days for our bodies to make the antibodies. During this time we can be very ill or even die.

When we have made an antibody we recover quite quickly. We then have a copy of the antibody that can be mass-produced when needed so that if the same pathogen invades our body again, we will be immune. This is called **active immunity**.

We can avoid catching many diseases by being immunised. This involves being injected with a harmless form of the disease that has antigens.

These antigens trigger our immune system to make antibodies. If we catch the harmful form of the disease in the future, the antibodies will protect us against it.

Sometimes when we catch a disease our bodies are not capable of making the antibody in time. This is a life-threatening situation. Fortunately, doctors can sometimes inject us with antibodies that have been made by someone else. This is called **passive immunity**.

Unfortunately these antibodies from someone else do not last. In order to gain long term protection we need to produce our own antibodies to the disease.

There are always risks when foreign substances are injected into our bodies. However, these risks are always much less than the risks of catching the real disease. This is why most parents are happy for their children to be immunised against several diseases such as measles and mumps.

 d Suggest why all children should be immunised even though there may be risks from the immunisation.

Fighting infection

Antibiotics are drugs that are used to treat bacterial and some fungal infections. In the past, antibiotics have been used too freely. Doctors have sometimes prescribed them to patients who have a cold, knowing that antibiotics do not work against viruses. Farmers have even used them to make their animals grow more quickly. This has resulted in some bacteria becoming resistant to antibiotics. We need to use antibiotics more carefully in the future.

The superbug, MRSA, is found in a lot of UK hospitals. The bacteria is dangerous because it is resistant to nearly all known antibiotics.

Scientists are attempting to find a new antibiotic that will be effective in killing the MRSA bacteria.

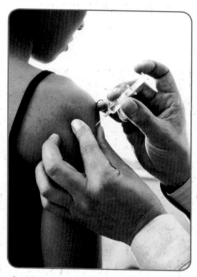

▲ *This person is being immunised*

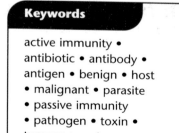

Drug testing

One of the jobs of scientists is to discover new drugs to protect us from disease. Discovering new drugs takes a long time. It is important to ensure that they are safe to use. First they are tested on animals and then on human tissue in the laboratory.

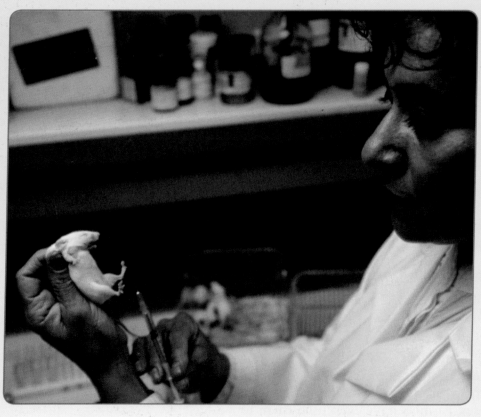

Some people object to testing drugs on animals and want to use other methods of testing such as computer modelling. Some scientists say that all drugs have side effects and it is much safer to test them on animals first.

Once the drugs are approved as safe then they are tested on humans to see how effective they are. This is done using 'blind' and 'double blind' drug trials.

Blind testing is when the patient does not know whether they are being given the real drug, or a fake pill made from substances such as flour and sugar.

This fake pill is called a placebo. The patient then has to say how effective the pill was. If those patients taking the real pill say it was effective and those patients taking the placebo say it was not effective, then the doctors know the pill is working.

Double blind testing is similar to blind testing, but this time even the doctor does not know which is the real pill and which is the placebo. Only an independent researcher knows which is which. This is so the doctor cannot give any hint to the patient about which might be the real pill.

Questions

1 Suggest why drugs are first tested on animals.

2 Why do you think some people object to testing drugs on animals?

3 Explain how blind testing of drugs works.

4 Explain why double blind testing of drugs is more reliable than blind testing.

Messages to the brain

In this item you will find out

- how the eye works and its problems

- how the brain keeps in touch with all parts of the body

- how different parts of the body communicate with each other

Imagine what it must be like to be completely blind. You would not be able to see all the colours and shapes around you. Your eyes send vast amounts of visual information to your brain.

Light enters through the eyeball where it is refracted by the cornea. The lens then focuses the light onto the retina.

The lens can change shape to focus light coming from different distances. This is called accommodation.

The lens can change shape because when the ciliary muscle contracts, it relaxes the suspensory ligaments that hold the lens, allowing the lens to become fatter. When the muscle relaxes it puts tension on the ligaments and stretches the lens to make it thinner.

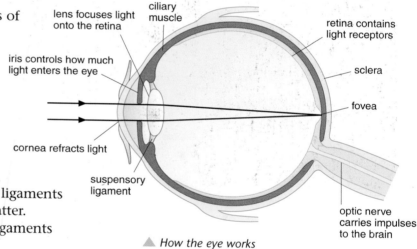

lens focuses light onto the retina

ciliary muscle

retina contains light receptors

iris controls how much light enters the eye

sclera

fovea

cornea refracts light

suspensory ligament

optic nerve carries impulses to the brain

▲ How the eye works

When we get older our eye accommodation slows down and becomes poorer because the lens gets harder. This means that senior citizens often have problems when they change from looking at something close to something far away, or the other way round.

Having two eyes enables us to judge distances quite accurately – try catching a ball with one eye closed! We can do this because the brain performs some clever mathematics. As an object approaches us, the eyes have to turn inwards. The brain can use this information to judge how far away the object is. This is called **binocular vision**.

TOBY MISUNDERSTOOD THE MEANING OF BINOCULAR VISION

a Explain how visual information reaches your brain through your eyes.

Colour vision

The retina contains specialised cells that can detect red, green and blue light. These three colours enable the brain to see the world in colour. Some people inherit a condition called **colour blindness.** It is caused by a lack of the colour-detecting cells in the retina. Some people are red/green colour blind. This means that red and green look the same to them.

▼ *Concave lenses correct short sight and and convex lenses correct long sight*

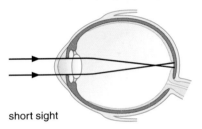

normal eye

Problems with sight

When the eye is working properly, the cornea and the lens focus rays of light onto the retina at the back of the eye. But sometimes the eyeball or the lens is the wrong shape.

In some eyes the light is focused short of the retina. This is called **short sight**. This can be corrected by using glasses or contact lenses with a concave lens. It can also be corrected by corneal surgery using a laser. In some eyes the light is not yet in focus by the time it reaches the retina. This is called **long sight**. This can be corrected using glasses or contact lenses with a convex lens.

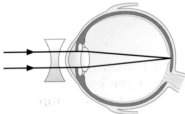

short sight

Concave lenses correct short sight.

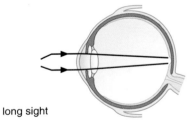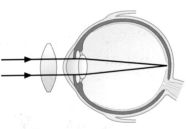

long sight

Convex lenes correct long sight.

b With reference to the lens, suggest why an eye may be short sighted.

The nervous system

Our nervous system consists of **sensory** nerves that carry information in the form of electrical impulses from the five senses to the brain and **motor** nerves that carry instructions to our muscles from the brain.

Each nerve consists of many nerve cells called **neurones**. Neurones are the longest cells in our body and can be over one metre in length. They act like telephone wires carrying information and instructions to and from the brain.

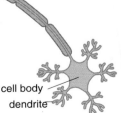

sheath   axon

a motor neurone carries instructions to our muscles

sheath   axon

a sensory neurone carries instructions to our muscles

dendrites

cell body

dendrite

▲ *This neurone carries information from our senses*

This neurone carries ▲ *instructions to our muscles*

The neurone is insulated by a **sheath**, which acts just like the plastic around the wires in an electric cable, stopping impulses travelling across from one neurone to another. The impulse travels down the **axon** from one end of the neurone to the other. The ends of each neurone have many branches called **dendrites**. This enables the neurone to connect with many other different neurones producing millions of different nerve pathways.

Unlike electric wires, neurones are not connected directly to other neurones. There is a gap between them called a **synapse**. The impulse crosses the synapse when it reaches the end of the neurone by triggering the release of a neuro-**transmitter** chemical into the synapse. The chemical diffuses across the synapse and binds with receptor molecules in the membrane of the next neurone. This causes the new neurone to produce an impulse that travels along it.

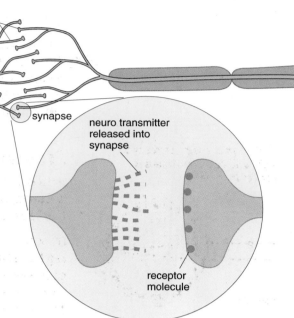

▲ *Synapses are the gaps between neurones*

 How is a neurone adapted to the job it does?

Reflex arc

It takes about a third of a second for an impulse from a sense organ to reach the brain and a return impulse to reach a muscle. If you have just picked up a very hot saucepan, this can be too long. Fortunately this time can be reduced by a **reflex arc**.

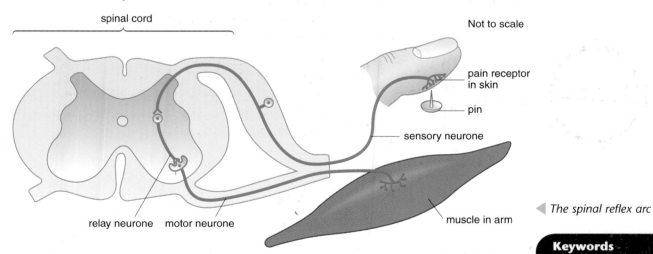

◀ *The spinal reflex arc*

In a reflex arc, the impulse goes from a **receptor**, along a sensory neurone into the spinal cord. As well as going up to the brain, a **relay** neurone connects directly to the motor neurone. The motor neurone goes to an **effector**, such as the muscle in the arm. This instructs the muscle to let go of the saucepan. By the time the brain receives the pain signal, the hand has already let go of the saucepan.

 Suggest why the reflex arc is so important to humans.

Keywords

axon • binocular vision • colour blindness • dendrite • effector • long sight • motor • neurone • receptor • reflex arc • relay • sensory • sheath • short sight • synapse • transmitter

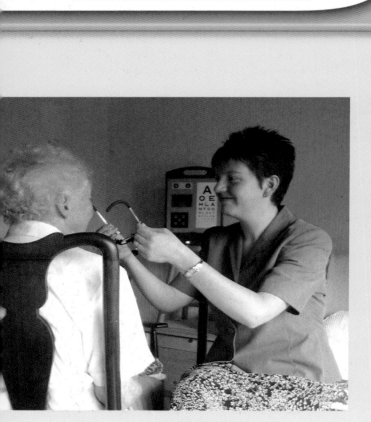

Seeing the optician

Margaret is a senior citizen. She has been having problems with her eyes so Helen the optician visits her in her home.

It is very important to have eye tests. Most people think that eye tests are performed to check whether you need to wear glasses. Eye tests, however, can tell the optician a lot about the health of the person.

The photograph below shows what Helen sees when she looks into a healthy eye.

The retina has a network of fine blood vessels.

Helen tests the pressure inside Margaret's eyeball. A blast of air is aimed at the front of the eye and the optician measures how far the cornea is pushed inwards. The higher the pressure in the eye, the less the cornea moves. If the pressure is too high, it is called glaucoma. Luckily Margaret isn't suffering from glaucoma.

Helen also checks Margaret for diabetes. Diabetes damages the blood vessels in the retina and it can lead to blindness. Unfortunately in the early stages there are no symptoms. This is why it is important that an optician spots the early signs of the disease. She can then refer the person to a doctor who will treat the patient for diabetes.

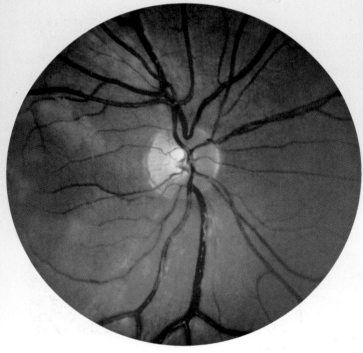

▲ Close-up of retina

Questions

1 Explain why it is important for people to have eye tests.

2 Explain why diabetes and glaucoma can damage the eye before someone realises that there is a problem.

3 Explain how Margaret is tested for glaucoma.

4 Suggest why Helen shines a bright light into Margaret's eye during an eye test.

Drugs make changes

In this item you will find out

- about different types of drugs and how they affect the body

- how drugs are classified by law

- about the effects of smoking and drinking alcohol

What do aspirin and heroin have in common? They are both painkilling **drugs**. You can buy aspirin in any chemist but if you are caught with heroin you could spend several years in prison.

So what are drugs? They are chemicals that produce changes within the body. The table shows the different types of drugs and the effects they have.

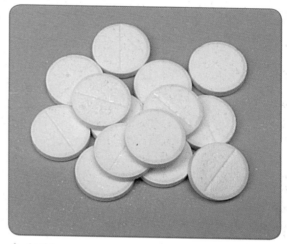

▲ *Aspirin*

Type of drug	Effect	Example
Depressants	slows the brain down	temazepam, alcohol, solvents
Stimulants	Increases brain activity	nicotine, ecstasy, caffeine
Pain killers	reduces pain	aspirin, heroin
Performance enhancers	improves athletic performance	anabolic steroids
Hallucinogens	changes what we see and hear	LSD, cannabis

 Suggest why many people drink coffee in the morning.

Depressants and stimulants act on the synapses of the nervous system. Depressants slow down transmission across synapses. Stimulants speed up transmission across synapses.

Some drugs are social drugs. They are called this because they can be taken legally and used for recreational purposes. Alcohol and tobacco are both examples of social drugs. Even social drugs can be addictive and cause harm.

▲ *Heroin*

Amazing fact

There are more synaptic pathways in the human brain than there are atoms in the universe.

Classification of drugs

The law classifies drugs into three different classes.

Class A

These are the most dangerous drugs such as heroin, cocaine, ecstasy and LSD. Illegal possession of this group carries the heaviest penalties with up to seven years in prison.

Class B

These drugs include amphetamines and barbiturates. Possession of drugs in this group can lead to up to five years in prison. Amphetamines such as 'speed' are stimulants while barbiturates are depressants.

Class C

These are mainly drugs prescribed by the doctor, and other drugs such as cannabis.

Smoking

Smoking is highly addictive because the tobacco in cigarettes contains a drug called **nicotine**. This is why it is so hard to stop once you start.

Cigarettes also contain tar and tiny particles called particulates. The tar is carcinogenic. It damages the lungs and can cause lung cancer, while the particulates can get trapped in the lungs and block the small airways.

When cigarettes are burned carbon monoxide is produced. This is a poisonous gas that prevents the blood from carrying oxygen and can lead to heart disease.

▲ A healthy lung (left), next to a smoker's lung (right)

b Look at the photograph of the healthy lung and the smoker's lung. Describe any differences you can see between them.

Effects of smoking

The trachea, bronchi and bronchioles are lined with mucus to trap dirt and microbes, and small hairs called **cilia** (ciliated epithelial cells). The job of the cilia is to waft mucus up from the lungs to the back of the throat where it is swallowed. Cigarette smoke contains chemicals that stop the cilia from working. This means that mucus accumulates in the lungs. The only way to get rid of it is to cough. This is often called a 'smoker's cough'. The build up of mucus in the bronchi can become infected and cause bronchitis.

c Describe how smoking cigarettes can result in a smoker's cough.

Look again at the photograph of the smoker's lung. You should notice that it has large holes. Smoking damages the air sacks in the lungs and produces spaces into which tissue fluid leaks. This is emphysema. It makes breathing very difficult. Chemicals in cigarette smoke can also cause cancer and heart disease. 90% of people who die from lung cancer are smokers.

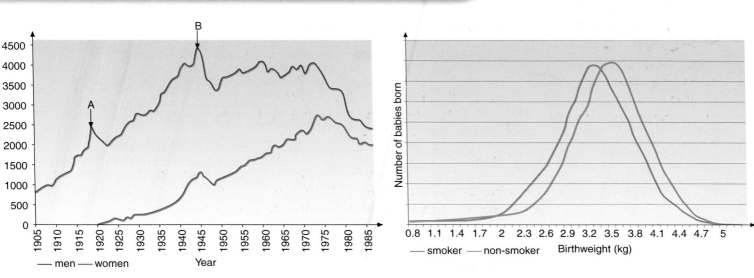

— men — women Year

— smoker — non-smoker Birthweight (kg)

d Look at the first graph. Suggest reasons for the two peaks, A and B.

e Look at the second graph. Explain the effect that smoking when pregnant has on the birth weight of babies.

Drinking alcohol

Alcohol is a poisonous drug that is removed from the body by the liver. Long term use can damage the liver causing a disease called **cirrhosis** of the liver. It can also damage the brain and nervous system. As the liver is responsible for a large number of the functions that take place in the body, liver damage is very serious and can lead to death. This is why long-term heavy drinking and binge drinking is so dangerous.

Alcohol consumption is measured in units. The drinks opposite all contain one or two units of alcohol each.

The recommended maximum weekly amount is 14 units for women and 21 units for men.

f If a person drank one pint of beer and two glasses of wine each day, how many units of alcohol would they drink in a week?

Alcohol also slows our reaction times. It increases the risk of having an accident when driving. This is why it is illegal to drink and drive.

beer
1 pint

2 units

wine

1 unit

cocktail

spirit

1 unit

2 units

▲ Each drink contains alcohol

Alcohol level in blood (mg/litre)	Reaction time compared with normal
0.8 the legal limit (two pints of beer)	4× slower
1.2 (3 pints of beer)	15× slower
1.6 (4 pints of beer)	30× slower

g What effect does drinking four pints of beer have on the chances of having an accident when driving a car?

Keywords

cilia • cirrhosis • depressant • drug • hallucinogen • nicotine • pain killer • performance enhancer • stimulant

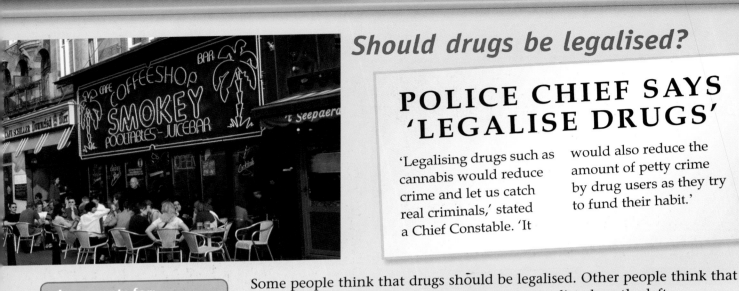

Should drugs be legalised?

POLICE CHIEF SAYS 'LEGALISE DRUGS'

'Legalising drugs such as cannabis would reduce crime and let us catch real criminals,' stated a Chief Constable. 'It would also reduce the amount of petty crime by drug users as they try to fund their habit.'

Some people think that drugs should be legalised. Other people think that drugs should not. Some of their arguments are listed on the left.

Arguments for:

People should be free to choose whether they take drugs or not.

Police spend too much time catching drug users.

Legalised drugs would be cheaper and safer for the users.

Cheaper drugs would mean less crime as users try to fund their habit.

Healthworkers would know who the users were and be able to help them kick the habit.

Free hypodermic needles would reduce the risk of drug users reusing them and passing on dangerous diseases such as AIDS.

Arguments against:

Increased drug use may lead to an increase in crime to pay for drugs.

People need to be protected against themselves.

Softer drugs such as cannabis may lead to using harder drugs such as heroin.

Drug use would increase if drugs were legalised.

Drugs affect reaction times. Road accidents and other types of accident would increase.

Cannabis abuse can lead to an increased risk of developing mental illnesses such as psychosis and schizophrenia.

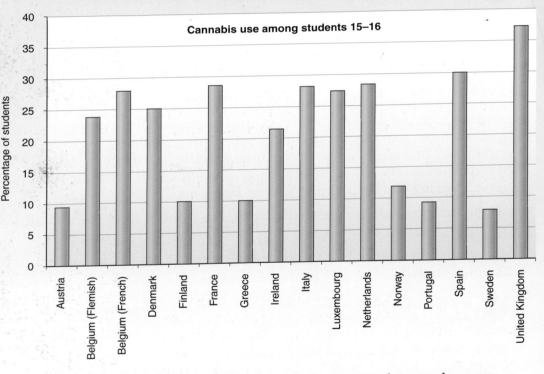

Cannabis use among students 15–16

(Percentage of students, by country: Austria, Belgium (Flemish), Belgium (French), Denmark, Finland, France, Greece, Ireland, Italy, Luxembourg, Netherlands, Norway, Portugal, Spain, Sweden, United Kingdom)

The Netherlands has a very liberal policy on drug use and many drugs are legalised. The graph shows drug use among students in different countries.

Questions

1 Some people think drugs such as cannabis should be legalised. Give reasons for and against this argument and explain your views.

2 Look at the graph. Explain what you think the effect of legalising drugs has had in the Netherlands.

3 Is it possible to draw a firm conclusion about what effect legalising drugs would have in the United Kingdom? Explain your answer.

LONGLEY PARK SIXTH FORM COLLEGE
HORNINGLOW ROAD
SHEFFIELD
S5 5SG

Bodies don't like change

In this item you will find out

- why keeping a constant internal environment in our bodies is important

- how the body maintains a constant temperature

- about some hormones and what they control

In the UK, temperatures can change from below freezing to nearly 30°C in summer. Your body has to work properly in an environment that is constantly changing.

In order for you to survive it is important that your body maintains a constant internal environment. This is called **homeostasis** and involves balancing what is taken into the body with what is given out. Your body has automatic control systems that keep temperature, water and carbon dioxide at constant levels so that cells can function at their optimum levels.

The reason why humans are so successful and have colonised every part of the planet is because of homeostasis. Most animals or plants are only found in specific areas. But humans are found everywhere.

Some animals that cannot control their own internal environment are often only found in parts of the world where conditions are just right. Others just shut down and go dormant during times when conditions are not to their liking. Even animals that can control their own internal environment, like the polar bear, are so adapted to their external environment that they can only be found in certain places in the world. All that fur would make it very hot for the polar bear if it lived at the equator. Unlike the polar bear, we can take our warm furry coats off.

▲ Clothing to protect against the sun

▲ Clothing in very cold weather

a Butterflies cannot control their temperature. It is always the same as their surroundings. Suggest why butterflies are not found at the North Pole.

b Explain homeostasis and give two examples of homeostasis in the human body.

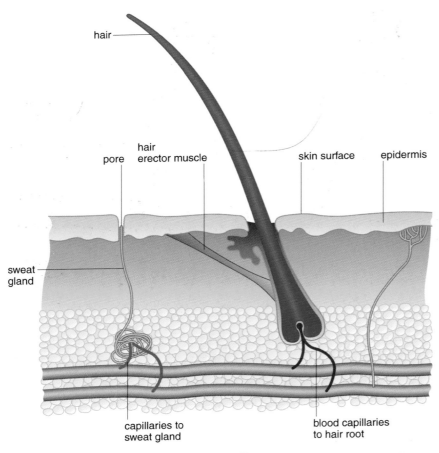

hair

hair
pore   erector muscle

sweat
gland

skin surface   epidermis

capillaries to
sweat gland

blood capillaries
to hair root

▲ A cross-section of the
skin showing the parts that
regulate temperature

Controlling body temperature

A healthy person has a constant body temperature of about 37 °C. The skin is the organ that is responsible for controlling this temperature.

This temperature control is important because the enzymes in our body work best at this temperature. Enzymes can be damaged permanently by a change in temperature. This is why a fever can be dangerous.

When we are too hot, more heat must be lost through the skin. We can do this in several ways. Blood vessels near the surface of the skin open. This is called **vasodilation**.

The skin becomes redder as warm blood is moved closer to the surface. The warm blood can then cool down as it loses heat to the surrounding air. We also start to sweat more. As the sweat evaporates, heat is removed from the skin and transferred to the environment.

When we are too cold, we must keep the heat inside our bodies. Blood vessels near the surface of the skin close. This is called **vasoconstriction**.

The skin becomes whiter as the warm blood is kept deeper inside our body. We stop sweating and start shivering. The muscles burn glucose and release heat energy into the body.

All of these changes are controlled by the brain which monitors the temperature of our blood. Because the outside temperature is changing all the time, the skin must maintain a delicate balancing act between heat lost and heat kept inside.

Temperature extremes

In extreme situations the body may be unable to control its temperature. If your body temperature gets too high, you can suffer from **heat stroke**. This can lead to **dehydration** and death if it is not treated. If your body temperature gets too low you can suffer from **hypothermia** which can also lead to death.

Sex hormones

The male and female **sex hormones** are responsible for the secondary sexual characteristics that happen at puberty. Hormones are chemicals used to transmit instructions around the body.

Amazing fact

Your skin has approximately 100 sweat glands per square centimetre.

Two of the female sex hormones are **oestrogen** and **progesterone**.

After a woman menstruates, oestrogen causes the lining of the uterus to thicken and re-grow new blood vessels.

Progesterone maintains the lining of the uterus. When the level of progesterone starts to fall towards the end of the monthly cycle, menstruation starts once more.

Oestrogen and progesterone together control other hormones produced by the pituitary gland that control the development and release of the ovum (egg) at **ovulation**.

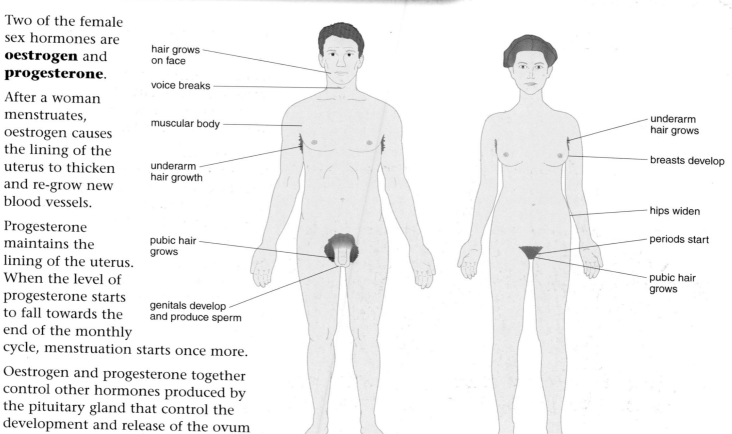

hair grows on face

voice breaks

muscular body

underarm hair growth

pubic hair grows

genitals develop and produce sperm

underarm hair grows

breasts develop

hips widen

periods start

pubic hair grows

▲ How sex hormones affect you at puberty

Controlling fertility

Hormones can be used for **contraception**. Drugs such as the contraceptive pill can be used to lower **fertility**. The drugs are similar to normal female hormones and prevent the body from ovulating and releasing eggs. No eggs – no babies!

Different female hormones can be used to treat women who are infertile because they do not produce enough eggs. The extra eggs produced can be donated to other women who cannot produce any eggs of their own.

Diabetes

Some people suffer from **diabetes**. They do not produce enough of the hormone **insulin**. Insulin converts excess sugar (glucose) in the blood into glycogen that is stored in the liver.

Diabetics have to be careful that they do not eat too much sweet food. They may also need to inject themselves with the hormone insulin to help them control their blood glucose levels. The dose of insulin that they inject will depend upon their diet and how active they are.

Keywords

contraception • dehydration • diabetes • fertility • heat stroke • homeostasis • hypothermia • insulin • oestrogen • ovulation • progesterone • sex hormone • vasoconstriction • vasodilation

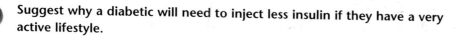

c **Suggest why a diabetic will need to inject less insulin if they have a very active lifestyle.**

Negative feedback

Lucy is doing research at university into negative feedback and its applications. During homeostasis, for example, negative feedback is the control system used to maintain a constant internal environment.

When we are very hot, the brain senses the temperature of the blood passing through it. It reduces the impulses along nerves to small muscles that surround capillaries in the skin, causing them to relax. This allows more blood to flow near the surface of the skin and to lose heat.

As the blood cools, the brain senses that the temperature of the blood passing through it is dropping. It then increases the impulses to the muscles, which causes them to contract and reduce the blood flow to the surface of the skin. Blood is diverted deeper into the body and heat is conserved.

Negative feedback is a process that can also be used by industry. The diagram belows shows how it can be used in sheet metal manufacture.

Gamma ray detectors measure the thickness of the sheet of metal. If the sheet of metal is too thick then the rollers move together. If the sheet is too thin then the rollers move apart.

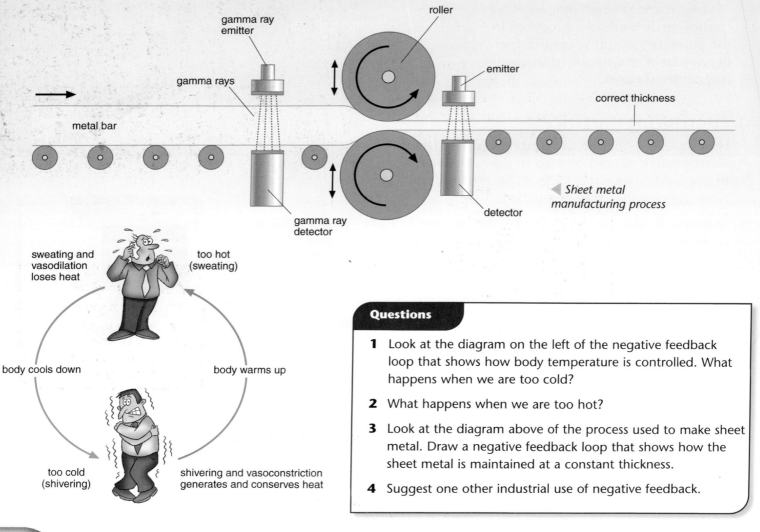

Sheet metal manufacturing process

Questions

1   Look at the diagram on the left of the negative feedback loop that shows how body temperature is controlled. What happens when we are too cold?

2   What happens when we are too hot?

3   Look at the diagram above of the process used to make sheet metal. Draw a negative feedback loop that shows how the sheet metal is maintained at a constant thickness.

4   Suggest one other industrial use of negative feedback.

It's all in the genes

In this item you will find out

- about the structure of DNA
- about chromosomes and genes
- why we are all different

DNA is an amazing chemical:

- it codes for all the information needed to make a new human being
- it can copy itself so that information can be passed on to future generations
- it is small enough to be stored inside the nucleus of nearly every cell in your body.

In order to fit all the DNA into each nucleus, the DNA is coiled many times to make it shorter. It is similar to a filament in a light bulb. A filament can be over 30 cm in length and coiling it up makes it shorter. The difference from DNA is that each coil is coiled again and again to make it very short indeed.

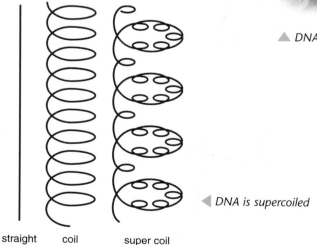

straight   coil   super coil

◀ *DNA is supercoiled*

▲ *DNA is very complex*

Amazing fact

If the information contained by one person's DNA was written down it would fill 1000 encyclopaedias.

 a   **Explain how such a long molecule as DNA can be squeezed into the nucleus of all our cells.**

The job of DNA is to control how your cells work by controlling the production of proteins. Some of these proteins are enzymes which cause different chemical reactions to happen in your body.

▲ *Rowan has more than 30,000 genes*

Chromosomes and genes

DNA is divided into areas called **genes**. A gene codes for a single instruction. A **chromosome** is made up of long, coiled molecules of DNA. All the chromosomes contain all the information for making a new human being.

In order to code for all the information, DNA uses four different chemicals, or **bases**. These four chemicals are called A, T, C and G. Just like using letters of the alphabet to make words and sentences, it is the order of these bases that stores the code or instructions for making a new human being. Each gene contains a different sequence of bases.

DNA only has four letters in its alphabet. This means that unlike words in English that are made up of about five or six letters, genes are made up of hundreds of bases. This makes chromosomes very long indeed. This is the four-letter **genetic code** that makes the hormone insulin.

atggccctgtggatgcgcctcctgcccctgctggcgctgctggccctctggggacctgacc
cagccgcagcctttgtgaaccaacacctgtgcggctcacacctggtggaagctctctacct
agtgtgcggggaacgaggcttcttctacacacccaagacccgccgggaggcagaggacc
tgcaggtggggcaggtggagctgggcggggccctggtgcaggcagcctgcagcccttg
gccctggagggggtccctgcagaagcgtggcattgtggaacaatgctgtaccagcatctgc
tccctctaccagctggagaactactgcaactag

Just imagine how long the sequence of bases would need to be to make baby Rowan, who has over 30,000 genes, many of which are much longer than the one for insulin.

b Explain the difference between a gene and a chromosome.

c Explain why (unlike words in English) the instructions needed to code for a gene are so long.

On or off?

Because different cells in your body use different proteins, they use different genes – the rest are switched off. For example, all nerve cells possess a complete copy of all your DNA, but they only need to use the genes that tell them how to be nerve cells. All the other genes (such as how to make haemoglobin or how to be a heart muscle cell) are not needed and are switched off.

d Explain why some genes are switched off.

Sexual reproduction

All humans have 23 pairs of chromosomes in most of their body cells (46 in total). Different species have different numbers of chromosomes, but they are always an even number.

e **Suggest why the number of chromosomes is always an even number.**

Human eggs and sperm (**gametes**) have half the number of chromosomes of body cells. They each have 23 single chromosomes.

When a sperm fertilises an egg during sexual reproduction the full set of 23 pairs of chromosomes is restored.

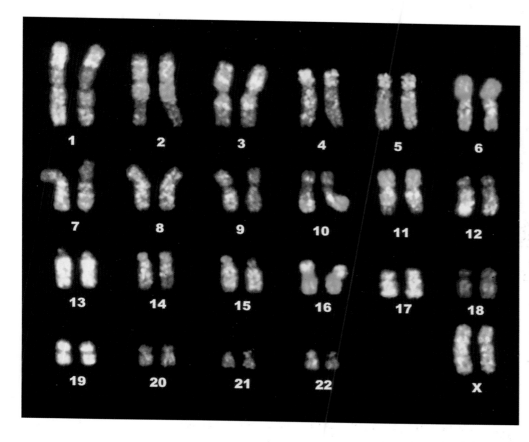

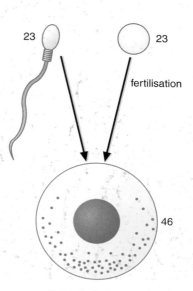

◀ *Human chromosomes*

This means that a baby contains the genetic material from both its parents.

Variety is the spice of life

The world is full of variation – apart from identical twins that have the same DNA, no two humans look the same. Some of this variation is caused by us inheriting different combinations of genes from our parents. Each gamete contains 23 chromosomes but it can be any one of the 23 from each pair. That is a lot of combinations.

More variation is caused at fertilisation itself, as any one of millions of sperm can fertilise an egg. It is hardly surprising that we are all different.

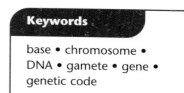

Keywords

base • chromosome •
DNA • gamete • gene •
genetic code

 f **Describe two ways that variation is brought about during sexual reproduction.**

 g **Explain why gametes have only 23 chromosomes when most other cells in your body have 46 (23 pairs).**

Why we are all different

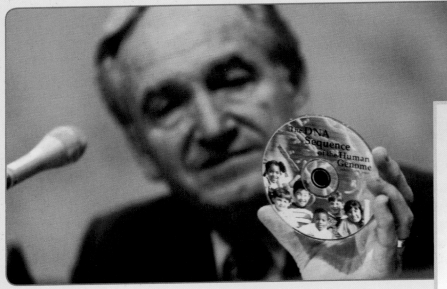

▲ *This CD holds all the data for making a new human being*

The Human Genome Project has been one of the great triumphs of science. It has been an international research effort to sequence and map all of the genes of members of our species, *Homo sapiens*. It was completed in April 2003 and, for the first time, gave us the ability to read nature's complete genetic blueprint for building a human being.

Benefits of mapping the human genome	Disadvantages of mapping the human genome
Scientists have a better understanding of genetic diseases.	Some people think it is against God or nature to alter the genetic code.
Possible cures may be found for many diseases.	New and dangerous diseases may be produced.
We will know if we are likely to develop a disease such as heart disease and cancer in later life.	Insurance companies may refuse insurance if they know that the person is at high risk of dying early.
Most scientists want the information to be freely available to everyone in the world.	Some people want to copyright some of the genes that they have discovered.

Questions

1 When was the human genome project completed?

2 List one advantage and one disadvantage in completing the project.

3 Suggest why some drug companies may want to copyright some of the genes.

4 Suggest why insurance companies may want to know about an individual's DNA.

Uniquely you

In this item you will find out

- how DNA can be altered

- how some diseases can be inherited

- about inherited characteristics

DNA is a very delicate chemical and can easily be damaged. Changes to DNA are called **mutations**. Mutations can change the sequence of bases in DNA or even remove some of the bases.

▲ *Our DNA can be altered*

Mutations are nearly always bad news. When they happen the message in the DNA becomes disrupted. Imagine you went through this book and took out some of the letters and replaced other letters with different ones. The book would very quickly become unreadable and useless. It is the same when mutations happen to DNA – the mutation prevents the production of the protein that the genes normally code for.

On some occasions the mutation can be useful. By pure chance, some changes do not produce gibberish but alter the message to make a different one. For example:

'Urgent message. Send more guns and ammunition.'

'Urgent message. Send more buns and ammunition.'

This can be very useful as it can produce even more variation within a species.

▼ *The stripy colours in the rose are caused by a mutated gene*

Mutations can be caused by many different things or they can even occur spontaneously. All of the following can cause mutations:

- ultraviolet light in sunshine or sun beds
- chemicals in cigarette smoke
- chemicals in the environment
- background radiation in the environment.

All of these factors can damage and change the sequence of bases in our DNA. Because we have so much spare unused DNA, the chances are that we will not notice most of these mutations. However, some mutations can cause diseases such as cancer.

Alleles

We know from the previous item that a baby's cells contain two complete sets of instructions, one from each of its parents. This means that there are two different versions of each gene called **alleles**. One allele comes from the mother and the other allele from the father. Obviously, the baby cannot use both of the alleles or it would end up with two of everything.

There are two different types of alleles: **dominant** alleles and **recessive** alleles. When someone has two alleles that are the same, they are called **homozygous**. If they have two different alleles they are called **heterozygous**.

Dominant alleles are the instructions that are used. Recessive alleles are only used if someone is homozygous and has two recessive alleles.

Breeding experiment

Breeding experiments can be done using pea seeds. Let us look at the gene for height in pea plants. It has two alleles. The recessive allele is for short plants and the dominant allele is for tall plants. We will use the capital letter **T** for the dominant tall and the lower case **t** for the recessive short.

If we cross a homozygous tall plant with a homozygous short plant, all of the offspring would be Tt. Because T is dominant for height, all the seedlings would grow tall.

This type of cross using just one character such as height is called a monohybrid cross.

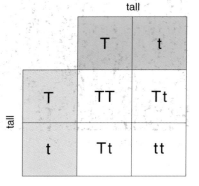

tall

short

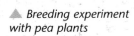

gametes

possible combinations

▲ Breeding experiment with pea plants

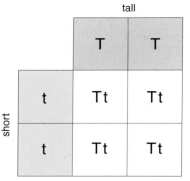

▶ A monohybrid cross

It is easier to see what is happening if we use a genetic diagram, such as the one above. This is called a punnet square.

Is it possible for two tall pea plants to produce seeds that will grow into short pea plants? The following genetic diagram on the left shows that it is.

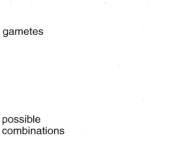

a Which of the seedlings in this diagram will be tall?

b Draw a punnet square to show what will happen when a plant with alleles Tt is crossed with a plant with alleles tt.

c What proportion of the plants will be tall?

▲ A cross between two heterozygous tall plants

Inherited diseases

Sometimes a mutation happens in the DNA found in one of the gametes. When this happens, the mutation can be passed on to the next generation. Fortunately such mutations are usually recessive. This is because if one parent donates a faulty allele, the allele from the other parent will probably not be damaged and can provide the correct instructions.

However, problems can arise if both parents carry the same faulty gene. The punnet square on the right shows what happens when both parents who are healthy, carry the faulty gene that causes cystic fibrosis.

One in four children will have the disease and half the children will carry the recessive allele (even though they are normal and healthy).

d   **Write down the combination of alleles that belong to the carriers of the disease.**

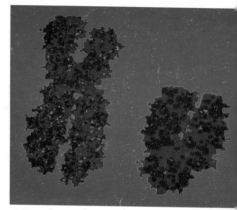

	mum	
	C	c
C	CC	Cc
c	Cc	cc

dad

C = the normal healthy allele

c = the faulty allele

Boy or girl?

Sex inheritance is controlled by a whole chromosome, rather than a single gene. In humans, males have two different chromosomes called X and Y. The X chromosome is larger than the Y chromosome.

Females have two chromosomes that are the same, called X and X.

This punnet square shows how sex is inherited.

 e   **What is the proportion of boys born compared to girls?**

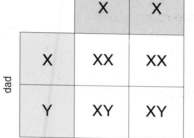

	mum	
	X	X
X	XX	XX
Y	XY	XY

dad

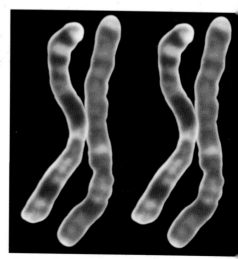

▲ *Male chromosomes XY*

▲ *Female chromosomes XX*

Variation

We have seen that mutation is a major cause of variation. From the breeding experiments we can see that fertilisation of gametes also leads to variation. When gametes are produced they all contain a different combination of chromosomes. These three factors ensure that nearly all human beings are very different from one another.

Genes or environment?

Scientists are now trying to determine the relative importance of our genes as opposed to the environment in making us who we are. We know that it is a combination of both, but not how much each one contributes. If we get the right genes we may be good at sport, good at school or very healthy. But we also know that to be good at sport requires hours of training, to be good at exams requires at lot of revision and to be healthy requires eating the right foods, taking exercise and not smoking.

Keywords

allele • dominant
• heterozygous •
homozygous • mutation •
recessive

To know or not to know?

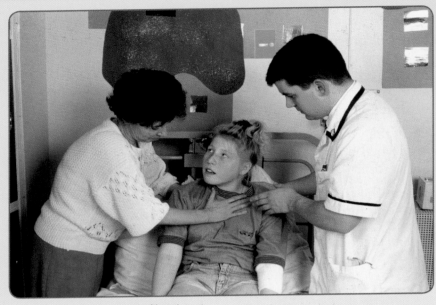

This girl inherited cystic fibrosis

As we learn more and more about genetics we also learn more and more about ourselves. Soon we will understand our genes so well that we will know our risks of getting heart disease or cancer. Genetic tests are already available to see if we carry the genes for various genetic diseases.

While all this knowledge can be very useful and enable us to make informed decisions about our lives, it can also have its disadvantages.

Examiner's tip

With questions based on choice you should always give both sides of the argument in your answer.

Benefits	Disadvantages
Knowing about our genes can enable us to decide whether or not to have children. If we knew that both our partner and ourselves were carriers for the cystic fibrosis gene, we would then know that there was a one in four risk of having a child with cystic fibrosis. We could then decide whether or not to have children or take some other course of action, such as having the gametes checked, to see if they were normal.	Society has not yet decided who owns the right to know about our personal DNA.
If we knew we had a high risk of dying from heart disease we could be more careful about our lifestyle and the food that we ate.	Some insurance companies are asking clients if they have ever had any DNA tests. If they find that the person is at risk they may refuse to insure that person or raise their premiums. This will enable them to make more money for their shareholders and keep premiums down for their other clients. This means that some people may be put off having tests to find out if they are at risk from genetic disease.

Questions

1 State two advantages of knowing about our own DNA.

2 State two disadvantages of knowing about our own DNA.

3 Explain whether you think insurance companies have the right to know about the genetics of their clients.

B1a

1   Finish the sentences by using words from the list:

age   diastolic   mmHg   systolic

Blood pressure is measured using
the units __(1)__.
Blood pressure varies according to __(2)__.
Blood pressure when the heart is contracting is
called __(3)__ pressure.   [3]

2   Which of the following word equations best describes
what happens during hard exercise?

A   glucose + oxygen → energy
B   glucose → lactic acid + carbon dioxide + energy
C   glucose → lactic acid + energy
D   glucose → water + lactic acid + energy   [1]

3   High blood pressure can be dangerous. Describe three
serious consequences that can result from having high
blood pressure.   [3]

4   During hard exercise, muscle fatigue can develop and
the body can build up an oxygen debt.

a   What is this type of respiration called?   [1]
b   Explain the cause of muscle fatigue.   [1]
c   Explain why the body builds up an oxygen debt. [1]
d   Explain how the body repays the oxygen debt.   [1]

5   Fitness can be measured in different ways. Each of the
following is a measure of fitness:

**strength   stamina   flexibility   agility   speed
cardiovascular efficiency**

Suggest and explain which type of fitness would be
required for each of the following activities:

a   marathon running   [1]
b   weight lifting   [1]
c   gymnastics   [1]

B1b

1   Finish the sentences by using words from the list:

amino acids   fatty acids   glycerol   simple sugars

Carbohydrates are made up of __(1)__.
Fats are made up of __(2)__ and __(3)__.
Proteins are made up of __(4)__.   [4]

2   Look back at the formula for calculating the
recommended daily allowance of protein on page 8.

a   Explain how the recommended daily allowance
of protein for a person is calculated.   [1]
b   Calculate the RDA of a person with a body mass
of 80 kg.   [1]
c   Suggest how the RDA may vary slightly with age.
    [1]

3   Describe how each of the following may affect a person's
diet:

a   vegetarian   **b**   self-esteem
c   religion   **d**   food allergy   [all 1]

4   Bile plays an important roll in digestion, even though it
does not contain any enzymes.

a   Which component of food does bile act upon?   [1]
b   Where is bile produced and stored?   [1]
c   Describe the function of bile on food.   [1]

B1c

1   Mosquitoes spread malaria.

a   What are organisms called that spread disease
from one organism to another?   [1]
b   The following statements are in the wrong order.
Write them out in the correct order.

A   The malarial parasite develops in the mosquito.
B   The person develops malaria.
C   A harmless mosquito sucks blood from an animal with
malaria.
D   A mosquito carrying the parasite sucks blood from
a healthy human.   [4]

2   Finish the sentences by using words from the list:

**active   antibiotics   antibodies   antigens
damage   passive   toxins**

Pathogens produce __(1)__ and can cause cell __(2)__.
Pathogens have __(3)__ which are locked onto by __(4)__.
After we recover from an infectious disease we often
have __(5)__ immunity to that disease.
Bacterial and fungal infections can also be treated
using __(6)__.   [6]

3   MRSA is sometimes called a 'superbug'. It is resistant to
almost all antibiotics. Scientists think that soon other
types of bacteria will become resistant to all antibiotics.

a   Explain how bacteria can develop resistance to
antibiotics.   [1]
b   Describe what can be done to prevent antibiotic
resistance in bacteria.   [1]

4   New drugs are tested before they are used on patients.

a   Explain why new drugs need to be tested.   [1]
b   Explain why double blind trials are used in drug
testing.   [3]

B1d

1   Explain the roll of each of the following parts of the eye:

a   cornea   **b**   iris   **c**   lens
d   retina   **e**   optic nerve   [all 1]

2 Look at the diagram of the reflex arc.

List an example of each of the following from the diagram:

a *stimulus*
b *sensor*
c *effector*

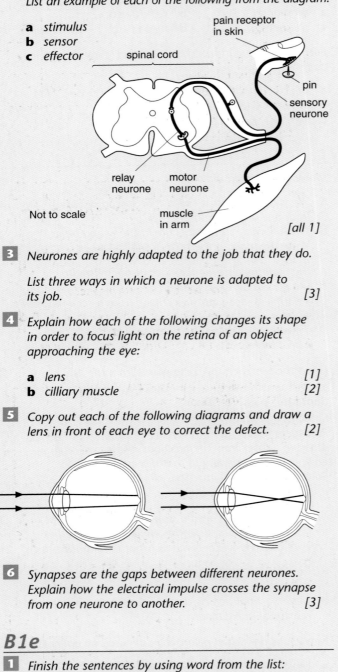

Not to scale

[all 1]

3 Neurones are highly adapted to the job that they do.

List three ways in which a neurone is adapted to its job. [3]

4 Explain how each of the following changes its shape in order to focus light on the retina of an object approaching the eye:

a *lens* [1]
b *cilliary muscle* [2]

5 Copy out each of the following diagrams and draw a lens in front of each eye to correct the defect. [2]

6 Synapses are the gaps between different neurones. Explain how the electrical impulse crosses the synapse from one neurone to another. [3]

B1e

1 Finish the sentences by using word from the list:
bronchi bronchitis cilia cough mucus
Cigarette smoke stops small hairs called __(1)__ from working.
The hairs normally propel sticky __(2)__ up from the __(3)__ to the back of the throat. A build up of sticky substance in the lungs causes __(4)__ and a smokers' __(5)__. [5]

2 Alcohol consumption is measured in units.

beer
1 pint
wine
spirit
cocktail
2 units   1 unit   1 unit   2 units

a How many units are there in three pints of beer? [1]
b If a person drinks three pints of beer a night, how many units will they consume in one week? [1]
c Men should not consume more than 21 units each week. If a man drinks three glasses of wine and one glass of whisky each day, will he be consuming too much alcohol? [1]
d Women should not consume more than 14 units each week. Suggest why the unit limit for woman is less than the unit limit for men. [1]

3 A synapse is the gap between two neurones. Explain the effect on a synapse of:

a *depressants*   **b** *stimulants*   [both 1]

4 Cigarette smoking causes tars and particulates to enter the lungs. Describe the effect of each of them on lung tissue. [2]

B1f

1 The body produces a hormone that controls the body's sugar level.
a Name this hormone. [1]
b What is the disease called when a person does not produce sufficient amounts of this hormone? [1]
c State two ways in which this disease can be treated. [1]

2 Finish the sentences using the following words:
close evaporates heat sweating open
When the body is too hot __(1)__ occurs, which __(2)__ and causes the body to cool. Blood vessels near the surface of the skin __(3)__ causing the skin to go red and radiate __(4)__ away. [4]

3 Oestrogen and testosterone are sometimes called the secondary sexual hormones.
a Explain what this means. [1]
b Describe the effects on a young teenager of:
i oestrogen   **ii** testosterone   [2]

4 The body maintains its internal environment using negative feedback.

a Explain how body temperature is controlled by negative feedback. [1]

b Give one example of how negative feedback can be used to control an industrial process. [1]

5 Explain the role of the following hormones on the menstrual cycle:

a oestrogen   b progesterone   [both 1]

6 Explain how female hormones can be used for:

a contraception   b fertility treatment   [both 1]

B1g

1 Chromosomes are found inside the nucleus.

a How many chromosomes are found in the nucleus of most human cells? [1]

b Is this number the same for all living organisms? [1]

c What is unusual about the number of chromosomes found in all living organisms? [1]

2 The information required to make a human being is coded in DNA.

a How many letters are in the DNA alphabet? [1]

b Explain how a complete set of DNA manages to fit inside the nucleus of a cell. [1]

3 Which of the following statements about gametes is true?

A Gametes contain the same number of chromosomes as other body cells.

B Gametes contain twice the number of chromosomes as other body cells.

C Gametes contain half the number of chromosomes as other body cells.

D Gametes do not contain any chromosomes. [2]

4 Explain what is meant by the term 'one gene, one enzyme'. [2]

5 Explain why some genes in some chromosomes are 'switched off' and are not used. [1]

6 Which of the following statements ensures that sexual reproduction always produces variation in the offspring:

A Offspring get DNA from both parents.

B Only one female ovum is released each month.

C Any one sperm can fertilise any one ovum.

D Sperm are much smaller than an ovum. [2]

7 Look at the following punnet square. It shows how sex is determined.

a Which are the sex chromosomes that determine males? [1]

b Which are the sex chromosomes that determine females? [1]

c Use the diagram to explain why equal numbers of boys and girls are born. [1]

	X	X
X	XX	XX
Y	XY	XY

B1h

1 Two people who could roll their tongue married and had children. Three of their children could roll their tongue and one could not.

Which condition is dominant and which is recessive? Explain your answer. [3]

2 Mutations are changes to the DNA in genes.

a Which of the following can cause mutations to DNA: water, radiation, chemicals, sound? [2]

b Explain whether most mutations are harmful or beneficial. [1]

3 Look at the following genetic cross between two tall pea plants.

T = Tall
t = short

a Which is dominant, tall or short? [1]

b Which pairs of alleles are homozygous? [1]

c What ratio of tall to short pea plants is produced by this cross? [1]

d Write down the allele for 'tall'. [1]

	T	t
T	TT	Tt
t	Tt	tt

4 The following cross shows the inheritance of a disease called cystic fibrosis.

a Explain why neither the mother or the father have the condition. [1]

b What proportion of children will have the condition? [1]

c Is the condition dominant or recessive? [1]

	C	c
C	CC	Cc
c	Cc	cc

5 It is now possible for parents to have genetic testing carried out so that they know the odds of having a child with a genetic disorder.

Explain the advantages and disadvantages of this testing. [2]

Dinosaurs are great, but why did they become extinct?

I guess they didn't look after their environment well enough and it changed so much they could not live there anymore.

No, it wasn't their fault. A giant asteroid crashed into the Earth and changed their environment for them. That's why they became extinct.

It seems strange that we are more intelligent than dinosaurs and yet we are destroying our environment but the dinosaurs didn't destroy theirs.

Maybe we are not so bright after all.

- Human beings share the planet with many other species of animals and plant and yet we are now doing enormous damage to all of our environments. This damage is causing global environmental change and it is occurring at a faster rate than it has done for millions of years.

- It is only by understanding our environment that we can learn how to take more care of it. We need to understand that the environment is not ours to do with as we want. We are simply looking after it for future generations. Your children will not thank us for destroying their inheritance.

What you need to know

- All living things are different and can be put into groups.

- Environmental, ecological and feeding relationships exist between different species of plant and animal.

- Plants are the source of all food and make it by the process of photosynthesis.

Pieces in a jigsaw

In this item you will find out

- about different ecosystems

- how data about ecosystems can be collected

- how to use keys to identify different animals and plants

We are very fortunate on Earth to have so many different types of **ecosystem**. An ecosystem is a place or habitat together with all the animals and plants that live there.

When scientists send probes to other planets, one landing site is probably much the same as any other. This makes it easier to understand and find out about the planet. Just imagine if aliens sent a probe to Earth and it landed in a desert. They might think that the whole of the planet was just one big desert.

We know more about the surface of the Moon than we know about some of the ecosystems on Earth. There are still many undiscovered **species** in places such as rainforests and the ocean depths.

Deserts, rainforests and meadows are examples of natural ecosystems. Artificial ecosystems are created by humans (examples include fields of a single crop, such as wheat or potatoes). Artificial ecosystems usually have a much lower **biodiversity**. This means there are fewer species living there.

Farmers usually have to use weed killers and pesticides to stop other species entering the ecosystem. They also usually use fertilisers to make sure the single crop species grows well before it is harvested.

▲ *Many strange species are waiting to be found at the bottom of the ocean*

▲ *Tractor spraying a field with fertiliser*

Using keys

It is important to be able to identify different animals and plants when studying an ecosystem. One way to do this would be to compare the **organism** with lots of different photographs of plants and animals to see which one it was. Unfortunately there are millions of different species, so you would need to look at millions of different photographs. This could take a long time.

 Estimate how long it would take to find a pupil at your school using a photograph, if you had to visit every class and look at every pupil.

A much better and faster way is to use a **key**. Keys work by dividing organisms into groups. Each group is then divided again and again. This may sound complicated but usually it only takes a few divisions to be able to identify an organism.

Try using this key to identify the following types of caddisfly larvae on the left.

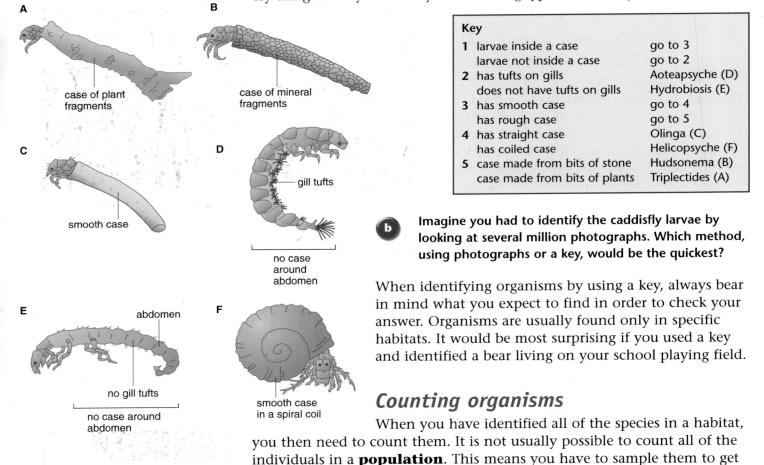

A — case of plant fragments

B — case of mineral fragments

C — smooth case

D — gill tufts / no case around abdomen

E — abdomen / no gill tufts / no case around abdomen

F — smooth case in a spiral coil

Key		
1	larvae inside a case	go to 3
	larvae not inside a case	go to 2
2	has tufts on gills	Aoteapsyche (D)
	does not have tufts on gills	Hydrobiosis (E)
3	has smooth case	go to 4
	has rough case	go to 5
4	has straight case	Olinga (C)
	has coiled case	Helicopsyche (F)
5	case made from bits of stone	Hudsonema (B)
	case made from bits of plants	Triplectides (A)

b **Imagine you had to identify the caddisfly larvae by looking at several million photographs. Which method, using photographs or a key, would be the quickest?**

When identifying organisms by using a key, always bear in mind what you expect to find in order to check your answer. Organisms are usually found only in specific habitats. It would be most surprising if you used a key and identified a bear living on your school playing field.

Counting organisms

When you have identified all of the species in a habitat, you then need to count them. It is not usually possible to count all of the individuals in a **population**. This means you have to sample them to get an idea of how big the population is.

If you did a survey on your school playing field, one way of collecting data would be to use a quadrat. A quadrat is a metal or plastic square that encloses 0.25 square metres. The quadrat is thrown at random and all the different species found within it are identified and counted. This is much easier than doing it for the whole playing field. It is then a simple matter of mutiplying the numbers counted by four to get the answer in square metres, and then multiplying the answer by the number of square metres in the playing field.

A much more accurate result can be obtained if several quadrat samples are taken and the average numbers are calculated. This is because not every quadrat will contain the same number of organisms.

 c If a student counts eight ladybirds inside one quadrat and the playing field is 12500 square metres, how many ladybirds would you expect to find in the playing field?

 d Not every quadrat will contain eight ladybirds. Suggest how you could make the results more reliable.

 e Suggest why this result will never be 100% accurate.

Quadrats cannot be used when sampling a pond or overgrown shrubland. Another method that can be used is 'capture and release'. A number of animals, such as beetles, are captured and marked with a tiny spot of paint. They are then released. Some time later, another sample of beetles is collected and examined to see how many are marked with the paint.

▲ *Student with a quadrat*

f Ten ladybirds are captured in a greenhouse. They are marked and then released. Sometime later, ten more ladybirds are captured and only one of them is marked. How many ladybirds do you think are living in the greenhouse?

g Suggest two different reasons why this method of counting populations may be unreliable.

◀ *A marked beetle*

Then and now

Jess is doing a school project on the changing environment. In the library she has found two maps.

The map on the left shows the land around Abbey Park in Leicester in 1826. The map is over 180 years old. The map on the right is up to date.

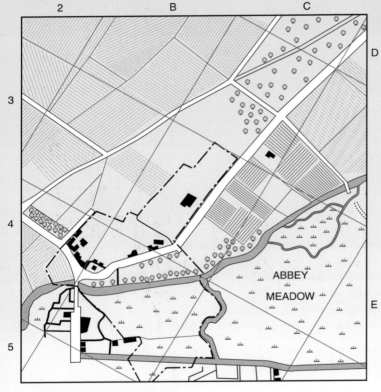

▲ The area 180 years ago

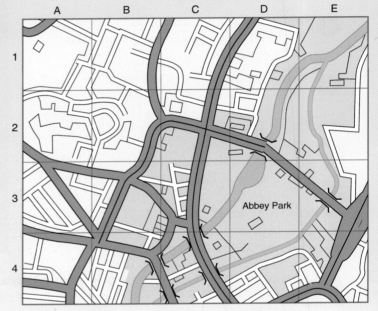

▲ The area today

Jess has also found some data on which species were around 180 years ago and which species are around today.

Questions

1 Describe two major changes that have taken place in the area.

2 What evidence suggests that the earlier map was not 100% accurate.

3 In the early map, the Abbey Meadows was a flood plain for the River Soar. List two species from the table that probably lived in the flood plain D3.

4 Suggest why the grid on the earlier map is in a different direction to the new map.

Map ref	Species	180 years ago	Today
D3	Bullrush	Lots	None
	Wheat	None	Lots
	Marsh warbler	Few	None
	Moorhen	Few	None
C4	Bluebells	Lots	Few
	Owls	Few	None
	Sparrows	Few	Lots
	Roses	None	Few
	Badgers	few	None
B3	Wheat	Lots	None
	Barley	Lots	None
	Field mouse	Lots	None
	Kestrel	None	Few

Pigeon-holing organisms

In this item you will find out

- what is meant by the word 'species'

- how living organisms are classified into different groups

- about the similarities and differences between species

▲ This frog is so new that it does not yet have a name

Scientists have just discovered a new species of frog. It is the smallest frog ever discovered in the southern hemisphere. It is only 1 cm long and can sit easily on a small coin.

A species is a group of organisms that reproduce with each other. For example, humans can only breed with other humans. This means that all humans belong to the same species called *Homo sapiens*.

Members of a species cannot usually reproduce with any other organism to produce offspring that are fertile. There are rare occasions when two organisms of different species can breed. They produce offspring called **hybrids**.

Animal hybrids are almost always infertile and cannot breed. Breeding a male donkey with a female horse produces a **mule**. The mule is sterile and is not a true species. This makes hybrids like the mule difficult to classify. It is not a donkey, nor is it a horse. It is a hybrid.

Just like us, a species has a first and a second name. The second name is often based on the name of the person who discovered it. This way of naming an organism is called the **binomial** system of classification.

◀ The mule is a hybrid

a Imagine that you have discovered a new species of frog. Its first name must be *Eleutherodactylus* but its second name can be anything you like.

Suggest a name for this new species.

I NAME THIS FROG FRED

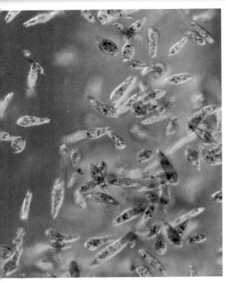

Plant or animal

Organisms are classified by placing them into different groups. In order to be able to do this, we need to know what criteria are used for each group. For example, animals and plants are classified into the plant or animal kingdoms using the following criteria.

It is an animal if it:	It is a plant if it:
can move independently	can only move in response to external conditions
cannot make its own food	makes its own food
is compact so that it can move about easily	is not compact and spreads out because it cannot move
does not have chloroplasts	has chloroplasts and is green

▲ *Euglena breaks the rules for animals and plants*

Rules and rule breakers

Unfortunately these rules do not always apply. Fungi cannot move and they are not green. This means we have to classify them in a group of their own as neither plant nor animal.

Some organisms have both plant and animal characteristics. A small microscopic pond organism called *Euglena* is green and can move. Animals can then be classified into two more groups:

- those with backbones – the **vertebrates**
- those without backbones – the **invertebrates**.

Vertebrates can then be placed into five different groups.

▲ *Worms do not have backbones and are invertebrates*

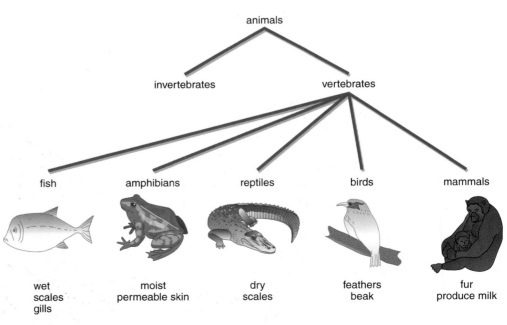

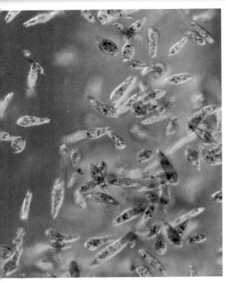

▲ *Snakes do have a backbone and are vertebrates*

▲ *Vertebrate classification tree*

Even with vertebrates, some species break the rules. The fossil of archaeopteryx has features of both birds and reptiles.

Scientists think that birds and reptiles are both descended from archaeopteryx. All of these five different groups can be divided into smaller and smaller groups until we get to individual species.

Similar habitats

If we look at similar habits in different parts of the world, we tend to find similar species living there. On grassland in England we find grazing animals like sheep, cows and horses. In a similar habitat, like the African bush, we also find grazing animals – but this time they are zebras and gazelles. They are different animals but they eat the same food and occupy the same ecological niche as cows and sheep.

Keywords

binomial • hybrid • invertebrate • mule • vertebrate

b Suggest why similar species evolve in similar habitats.

c Suggest what grazing species have evolved in the Australian outback.

Similar species

The apes, for example, consist of many closely-related species. Although they are all similar, they are also different from one another. The gorillas are big and heavy because they have evolved to live and gather food on the ground. Chimpanzees are smaller and lighter because they have evolved to gather food up in the trees.

Gorillas and chimpanzees have both evolved from a common ancestor in the recent evolutionary past.

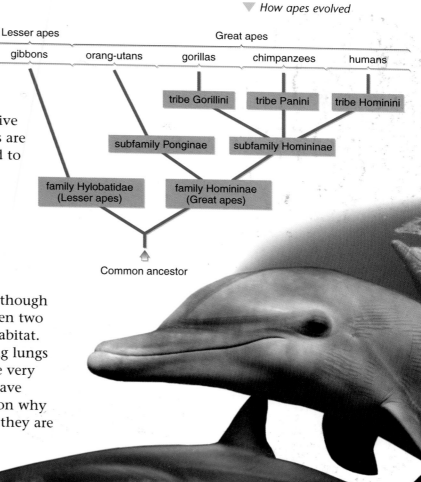

▼ How apes evolved

Dolphins and sharks

Some species are similar to one another even though they are not closely related. This happens when two very different species are living in the same habitat. Dolphins are mammals. They breathe air using lungs and produce milk for their offspring. They are very similar to sharks, but sharks are fish. Sharks have gills and do not feed their offspring. The reason why dolphins and sharks are similar, even though they are not related, is because they both live in the same habitat. They have to swim in water so they both have to be streamlined and have fins.

▶ Dolphins and sharks have many similarities

From chaos into order

Eric and Sam are on a school field trip to one of the streams near their school. They are using nets to catch the insects found in the stream. Once they catch them they make notes and drawings of their features and then put the insects back. When they are back in the classroom they classify the insects.

To do this, they divide the insects into two groups. They then take one of those groups and divide into two more groups. Eric and Sam repeat the process until all the insects are classified. They know that there are many different ways of putting the insects into different groups. When we classify living organisms, we try to choose those organisms that are closely related in terms of their evolution and put them into the same group.

Questions

1 What choice could Eric and Sam make when they divide them into the first two groups?

2 How many times can they divide them into groups until they are all classified?

3 Can you think of any other way of classifying the insects. What would be your first choice this time?

4 Suggest why classifying flowers based on their colour is not as good as classifying them based on the structure of the flower.

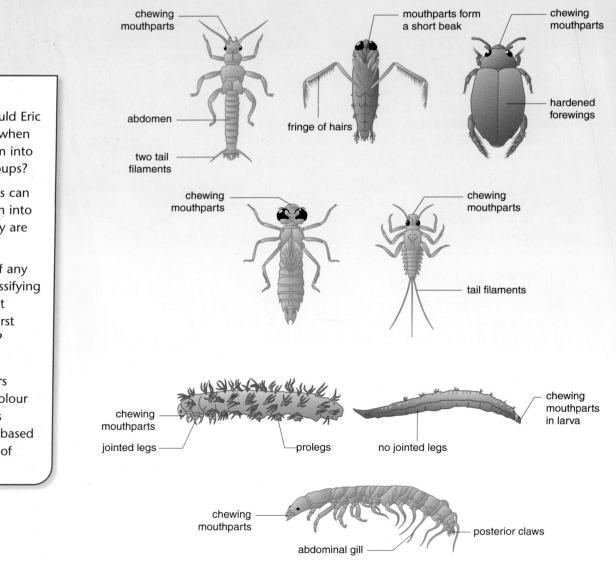

Plant magic

In this item you will find out

- about photosynthesis and how glucose can be converted into other substances

- how the rate of photosynthesis can be increased and the effect of limiting factors

- about the relationship between photosynthesis and respiration

Most students know that plants make food by **photosynthesis** from **carbon dioxide** taken from the air. It is hard to imagine that trees absorb tonnes of carbon dioxide when they photosynthesise and that millions of tonnes of carbon dioxide are put into the atmosphere when we burn fuel.

Even when we think about it, it is still difficult for us to imagine that wood (and all the vegetables and fruits that we eat) are made by plants, from carbon dioxide taken from the air and water taken from the soil.

 a Which human activities put carbon dioxide back into the air?

Plants make food by converting carbon dioxide and water into **glucose** and oxygen. To do this they need energy from sunlight and a green chemical called chlorophyll.

$$\text{carbon dioxide} + \text{water} \xrightarrow[\text{chlorophyll}]{\text{light energy}} \text{glucose} + \text{oxygen}$$

$$6CO_2 + 6H_2O \xrightarrow[\text{chlorophyll}]{\text{light energy}} C_6H_{12}O_6 + 6O_2$$

b Two and a half acres of forest absorbs 1 tonne of carbon dioxide a year. One gallon of petrol produces over 8 kg of carbon dioxide. If a motorist uses 500 gallons of petrol a year, how many acres of forest are needed to absorb all of the carbon dioxide?

Because it is the leaves that carry out photosynthesis and absorb the carbon dioxide from the air, forests are sometimes called 'food factories'.

Amazing fact

One hectare of corn produces enough oxygen by photosynthesis for about 325 people.

Examiner's tip

The equation for photosynthesis is the same as the equation for aerobic respiration backwards.

Flour is mainly starch

Glucose

Glucose is very soluble and is dissolved in the plant's sap. The plant can then transport the dissolved glucose to any other part of the plant. Very often it is transported to the plant's roots for storage. Glucose is a simple sugar and is used in sweets such as chocolate bars.

Plants use the glucose for **respiration**. This releases energy for the plants to use.

Uses for glucose

Plants make much more glucose than they need for respiration. They cannot store the glucose because it is so soluble. Some plants convert the glucose into insoluble **starch**, **fats** or **oil** for storage.

When starch, fat or oil is placed inside a cell, it cannot get out because it is insoluble. Plants that store food as starch include the potato, wheat and corn. Bread flour is mainly starch.

Plants, such as oil seed rape, convert and store the glucose as oil or fat.

Glucose is also converted into **proteins** and **cellulose**. Proteins are used for growth and repair of the plant. Cellulose is the material that plant cell walls are made from.

Humans cannot digest cellulose and when we eat plants, the cellulose forms part of the roughage that passes straight through our gut.

Oil seed rape

Making photosynthesis work faster

We can make photosynthesis work faster – which is useful because it increases the rate at which farmers can grow food.

- We can increase the amount of carbon dioxide for the plants to use. We can only do this effectively in a greenhouse because carbon dioxide is a gas, Outside, the carbon dioxide would blow away.
- We can increase the amount of light reaching the plants. We can either increase the brightness of the light or the number of daylight hours.
- We can increase the temperature around the plants. Photosynthesis is a chemical reaction. Chemical reactions are faster and happen more often when the temperature increases.

 Suggest how farmers could increase the levels of carbon dioxide in a greenhouse.

Limiting factors

Carbon dioxide, light and temperature are all **limiting factors**. A lack of any one of them can limit how fast photosynthesis can go.

Look at the graphs. As light intensity increases, so does the rate of photosynthesis, but then it reaches a maximum and the graph levels out. Increasing the light further will not increase the rate of photosynthesis. But if the level of carbon dioxide is increased, then the rate of photosynthesis increases once more. The level of carbon dioxide is a limiting factor.

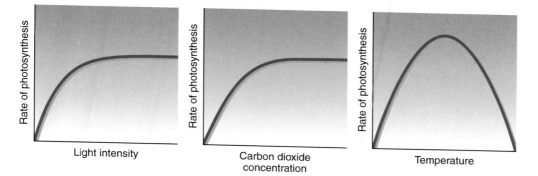

 Carbon dioxide is a limiting factor on the rate of photosynthesis

d What happens to the rate of photosynthesis as the amount of light increases?

e What happens to the rate of photosynthesis as the amount of carbon dioxide increases?

Respiration versus photosynthesis

Plants respire all of the time. This means that they constantly use oxygen and release carbon dioxide. During the hours of daylight they also photosynthesise.

Plants photosynthesise much faster than they respire. This means that during the day they release much more oxygen than they use and absorb much more carbon dioxide than they release. So over a 24-hour period they produce much more glucose by photosynthesis than they use for respiration.

f Look at the graph on the right. How many times each day is the rate of photosynthesis equal to the rate of respiration?

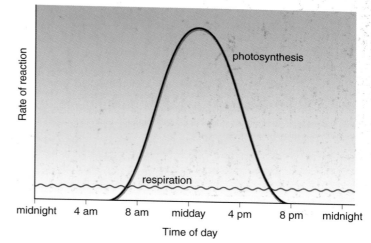

 Respiration and photosynthesis

Keywords

carbon dioxide •
cellulose • fat • glucose
• limiting factor • oil •
photosynthesis • protein •
respiration • starch

Increasing food production

Many commercial tomato growers use greenhouses. Modern greenhouses are very complicated places and are often controlled by a computer.

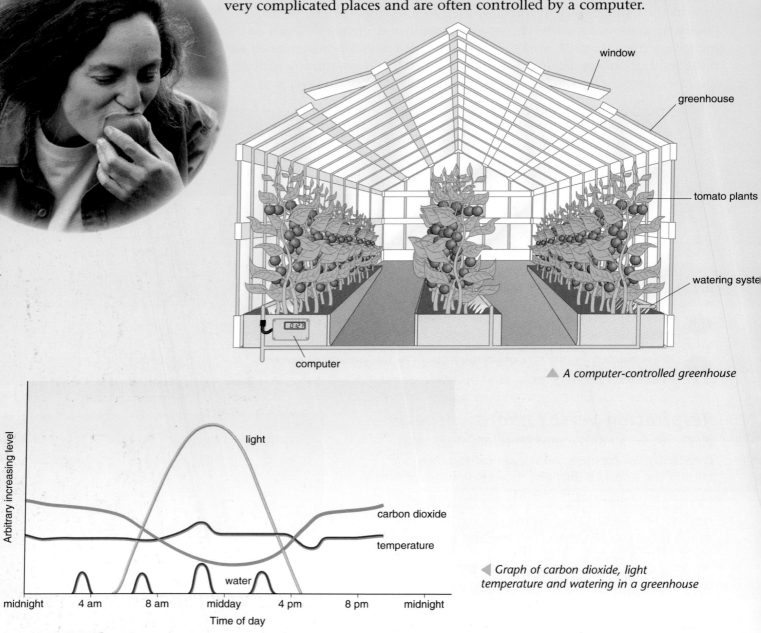

▲ A computer-controlled greenhouse

◀ Graph of carbon dioxide, light temperature and watering in a greenhouse

Questions

1 Suggest at what time of day the computer opened the windows.

2 State how many times during the day the computer turned on the watering system.

3 Suggest why the carbon dioxide levels fell during the hours of daylight.

4 Using only information from the graph, suggest two ways that the farmer could increase his crop of tomatoes.

5 Suggest why tomato crops grown in a greenhouse tend to be much taller than tomato crops grown outdoors.

The fight for survival

In this item you will find out

- how different animals and plants compete with each other

- about predators and prey

- about organisms that rely on other species for their survival

Each individual organism on the planet is in **competition** with all the other organisms for survival. Organisms that fail to compete successfully will die. Only successful organisms will go on to survive and breed.

Plants and animals do not just compete for food. They also compete for water, shelter, light and minerals. Because the availability of these factors varies from place to place, the distribution of organisms is also affected. For example, only organisms that can survive on small amounts of water, such as the camel, are found in the desert.

In the western world, we sometimes forget about this competition for survival because we can now control so many aspects of our own environment. Many of us have warm homes and can buy our food from shops, and now only compete for a better job or lifestyle.

Competition ensures that the population of one species does not get too big. When the population does get too big, something always happens to bring it back down to its usual size. This may be shortage of food or water, lack of space or shelter, or disease.

Sometimes when there is less competition for food, a population can explode in numbers. This happens when swarms of locusts breed and then feed on crops. Eventually the swarm runs out of food and begins to starve.

▲ In more wealthy countries humans compete for status and possessions

▲ Locusts do a lot of damage to crops

▲ Red squirrel

▲ Grey squirrel

Competition between species

Habitats can only support so many species, particularly if they are similar and competing for the same food and space. The most successful species survives and the least successful dies. This has happened to the red squirrel. When the grey squirrel was introduced to this country, it was so successful that it out-competed the native red squirrel. Red squirrels are now only found in a few isolated places in the United Kingdom. You are very lucky if you have ever seen a red squirrel in the UK.

a Suggest why the grey squirrel is more successful than the red squirrel.

b Suggest how we can ensure that populations of red squirrel continue to survive.

When two different species compete for the same part of a habitat, scientists say that they are competing for the same **ecological niche**.

Another example of an introduced species winning the battle for an ecological niche is when mink escaped from mink farms. The mink soon became wild and out-competed the native otter.

Predators and prey

Predators and **prey** both affect the size of each other's population. When the population of the predator increases, they eat more prey. This makes the prey population fall. Because there is now less food, the population of the predator falls. There are now fewer predators so the population of the prey increases again. The cycle of predator and prey populations increasing and falling is repeated over and over again. This relationship helps to maintain the balance in numbers and stops any population increase exploding out of control.

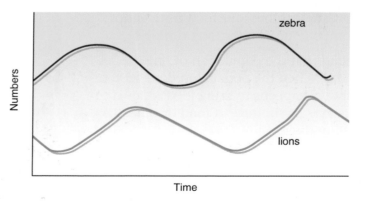

▲ Predator and prey populations

c Look at the graph. What do you notice about the numbers of predators compared with the numbers of prey?

d Describe one other pattern that you can see in the graph.

Parasites

Some organisms are so competitive that they can only survive by living on or in the body of other organisms. They are called parasites and the organisms they live on or in are called hosts. Examples of parasites include fleas and tapeworms. Fleas survive by living on the skin of an animal and sucking its blood. This can weaken the animal and also introduce dangerous diseases into the bloodstream.

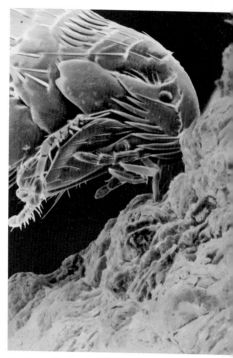

▲ Fleas feed off blood

Tapeworms grow in the gut of an animal and feed off the food the animal eats. In severe cases they can cause a blockage in the animal's gut. They can grow up to several metres long.

Parasites always live at the expense of the host organism. They never do any good and usually harm the host. This is not always a sensible way of competing. If the host dies, the parasite will often die as well.

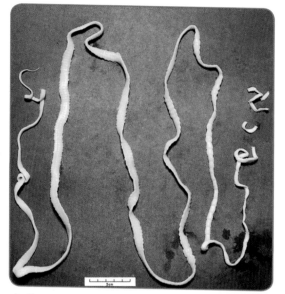

▲ Tapeworms feed off other animals' food

e **Why it is not a good idea for the parasite to kill the host on which it lives?**

▼ Oxpecker birds and giraffes live together

Mutualism

Some organisms live together. This is called **mutualism** and both organisms benefit from the relationship. One example is the oxpecker bird which eats small parasites from the fur of mammals.

Another example is nitrogen fixing bacteria that live in the root nodules of leguminous plants, such as peas, bean and clover. The plant provides the bacteria with sugars for food. The bacteria convert atmospheric nitrogen into a form that the plants can use to make protein.

This close interdependence ensures that when one organism survives and is successful so will the other one. It also determines the distribution and abundance of different organisms in different habitats.

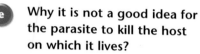

Keywords

competition • ecological niche • mutualism • predator • prey

Illegal immigrants

▲ Harlequin ladybird

Britain's best-loved beetle, the ladybird, is under threat from the world's most invasive ladybird species – the harlequin ladybird.

Originally from Asia, the harlequin ladybird was first spotted in the UK in September 2004. Since then many sightings have been reported, but these have mainly been confined to the south east of Britain.

There are 46 species of ladybird in Britain and the harlequin ladybird is a potential threat to the survival of all of them. It is an extremely voracious predator that easily out-competes native ladybirds for food. When their preferred food of greenfly and scale insects is not available, the harlequin readily preys on native ladybirds and other insects such as butterfly eggs, caterpillars and lacewing larvae.

It was introduced into many countries as a biological control agent against aphid infestations in greenhouses, crops and gardens. Populations have been found in North America, France, Germany, Luxembourg, Belgium, Holland, Greece and Egypt. In France, Belgium and Holland numbers are increasing every year.

They can disperse rapidly over long distances and so have the potential to spread to all parts of the United Kingdom.

> **Examiner's tip**
>
> **Don't be put off by unfamiliar examples in the examination. The principles are the same.**

Key
- ● 2004
- ● 2005

▲ Distribution of harlequin ladybirds (*H. axyridis*)

Questions

1 Suggest why more sightings of the harlequin ladybird have occurred in the south east corner of the UK.

2 In what way does the harlequin ladybird out-compete the native ladybird?

3 Explain the term 'biological control'.

4 Where did the harlequin ladybird originate?

5 Suggest how the ladybird can move rapidly over long distances.

Adapt or die

In this item you will find out

- how some organisms manage to survive in harsh environments
- about different kinds of adaptations

▲ Spider on web

Usually the environment changes very slowly. Over millions of years, the land we called England has been under a warm tropical ocean, had lush tropical rainforests, been covered in ice and even been a desert.

The changes take place so slowly that we do not usually notice them over the course of a lifetime. This means that animals and plants have a long time to **adapt** to the environment when it changes. Plants and animals adapt to their surroundings in order to survive. Adaptations help them compete for limited resources and increase in number.

 a   **Suggest other ways that a spider is adapted to its environment.**

Pythons are adapted so they can swallow prey that is larger than their own heads.

Spiders make sticky webs to catch prey, but the spider is adapted so that it does not stick to its own web.

When the environment changes very quickly, organisms do not have time to adapt. Some scientists think this is what may have happened to the dinosaurs.

▲ Python swallowing prey

 b   **Explain why the dinosaurs became extinct when they were so perfectly adapted to their environment.**

It has been suggested that a large meteorite impacted with the Earth near Mexico. An explosion like this would have caused such rapid change to the Earth's environments that the dinosaurs would not have had time to adapt.

The more quickly that animals and plants can adapt, the more successful they will be. Their populations will have large numbers because there is less competition from other animals and plants.

How they adapt will also affect their distribution. For example, animals and plants that have adapted to live in dry environments will tend to be found in deserts.

IF DINOSAURS HAD BEEN ABLE TO ADAPT

The polar bear

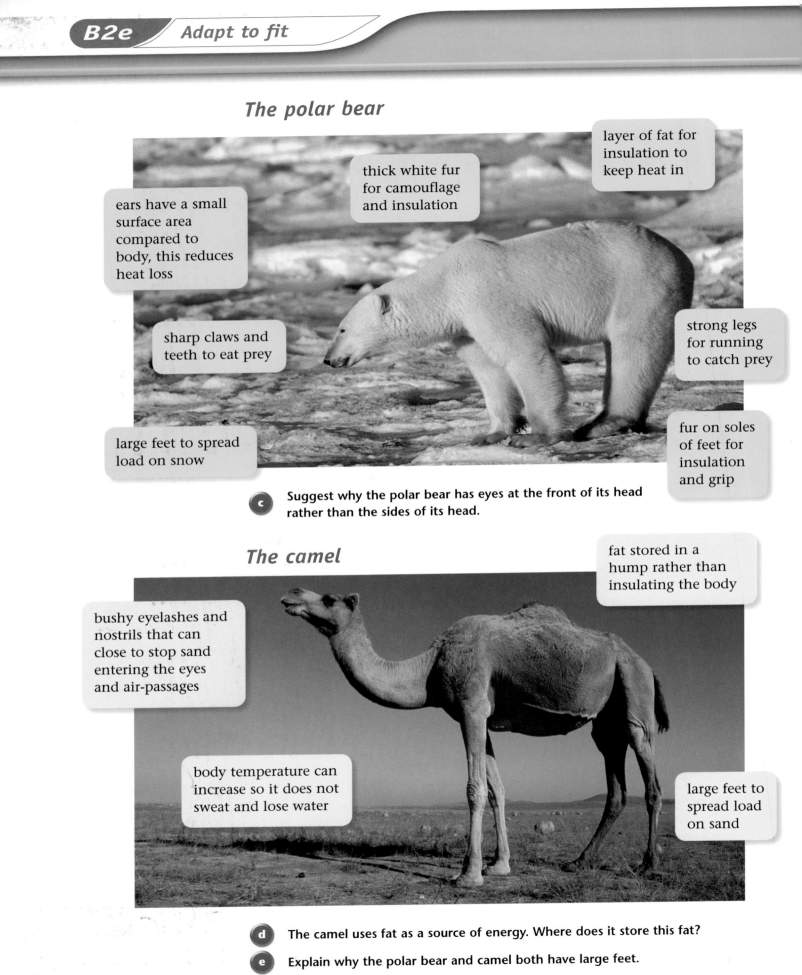

layer of fat for insulation to keep heat in

thick white fur for camouflage and insulation

ears have a small surface area compared to body, this reduces heat loss

sharp claws and teeth to eat prey

strong legs for running to catch prey

large feet to spread load on snow

fur on soles of feet for insulation and grip

c Suggest why the polar bear has eyes at the front of its head rather than the sides of its head.

The camel

fat stored in a hump rather than insulating the body

bushy eyelashes and nostrils that can close to stop sand entering the eyes and air-passages

body temperature can increase so it does not sweat and lose water

large feet to spread load on sand

d The camel uses fat as a source of energy. Where does it store this fat?

e Explain why the polar bear and camel both have large feet.

The cactus

thick cuticle to reduce water loss

leaves reduced to spines to prevent water loss

green stem for photosynthesis

thick, round fleshy tissue to reduce surface area to volume ratio – to reduce water loss and store water

long roots to reach water

 Cacti also have lots of surface roots that spread out over a large area. Suggest why this is a useful adaptation in a dry desert.

Wind pollination

feathery stigmas to catch small, light pollen

Insect pollination

colourful petals to attract insects

sweet nectar to attract insects

sticky pollen to stick to insects

 Suggest why wind pollinated flowers do not have brightly coloured petals.

Keyword

adapt

Taking adaptation to extremes

Some organisms are so adapted to their environment that they live on the very edge of survival. One such place is deep under the ocean near 'black smokers'.

▲ Black smoker

Black smokers are deep sea, underwater hydrothermal vents, found in areas of undersea volcanic activity, which release large amounts of superheated solutions of minerals into the ocean. As the hot steam and water meet the cold ocean, the minerals crystallise out and form rocky deposits.

The minerals form into tall black towers that support a large variety of life such as tube-worms, giant clams and barnacles. No sunlight reaches these great depths. The organisms live in total darkness and plants cannot photosynthesise.

The energy to support this food chain comes from chemosynthetic bacteria that live on the very edge of the superheated water. They obtain their energy by chemosynthesis as they metabolise the sulfur in the hot springs. Some scientists think that this may be where life on Earth originally began.

Questions

1 Explain why green plants such as algae are not found deep in the ocean.

2 What organism provides the energy for all life found close to black smokers?

3 Explain how this organism provides energy for the food chain.

4 Suggest why it was difficult for life to have evolved in such a hostile place.

5 Suggest why scientists are so interested in life that has evolved near black smokers.

All change

In this item you will find out

- how fossils are formed and how they can be useful in understanding evolution

- what happens when environments change

- about Darwin's theory of natural selection

Humans have always asked the question 'where do we come from and how did we get here?' Over the centuries, different people have had different theories. Some people, called creationists, believe that their God created all life and that organisms were placed on Earth ready made. For example, buttercups were created as buttercups and humans created as humans.

Most scientists now think that the clues to our origin lie in the **fossil** record. However, creationists believe that even the fossil record was created by God. Scientists think that the fossil record provides evidence for the evolution of species over a period of millions of years (long before creationists believe their God created the world). The fossil that most people are familiar with and have seen for themselves is the ammonite, which scientists think existed 500 million years ago.

Sometimes when an organism dies, it becomes covered with sediment. Over many years, hard parts (such as shells and tough leaves) are gradually replaced by minerals which form the fossil.

On rare occasions, the whole organism may be preserved. This happens when bacteria that use oxygen are prevented from making the organism decay. Examples include insects embedded in amber, dinosaurs that fell into peat bogs or tar pits or mammoths that became frozen in ice.

The fossil record is not complete. This is because fossils of most organisms have not yet been discovered. Fossilisation also only very rarely occurs – most organisms do not form fossils when they die. Soft tissue usually decays and does not fossilise.

 State one piece of scientific evidence that supports evolution and one piece of scientific evidence that supports creationism.

▲ Ammonite fossil

Natural selection

Organisms that are better adapted to their environments are more likely to survive. When an environment changes, many of the plant and animal species living in that environment survive or evolve, but many become extinct because they cannot change quickly enough.

In 1859, Charles **Darwin** put forward his theory of **natural selection**. Members of the same species are different from each other. This is called natural variation. These members are in competition with each other for limited resources. Because all organisms are slightly different, some are better adapted to the environment than others.

These organisms are more likely to survive and breed. This is called survival of the fittest. The offspring inherit these successful adaptations. The organisms which do not carry these characteristics die out because they are not able to compete successfully.

We now know that adaptations are controlled by genes and that these genes are inherited by the next generation.

▲ Charles Darwin

Evolution of the horse

In the 1870s, the palaeontologist O.C. Marsh published a description of three fossil horses that he had found in North America. Marsh thought that the horse had evolved through a series of stages from one form to another until they had evolved into the modern horse. This was called straight line evolution.

As more fossils of horses were discovered it became clear that the evolution of the horse was much more complicated than Marsh originally thought. The three fossil horses that Marsh discovered were Orohippus, Miohippus and Hipparion.

Marsh noticed that the legs of horses had evolved to allow them to run very fast.

Early horses ran on all four fingers (early horses did not have a thumb). Modern horses run on a hoof, which has evolved from just the middle finger.

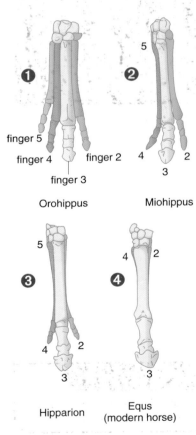

finger 5
finger 4   finger 2
finger 3

Orohippus   Miohippus

Hipparion   Equs (modern horse)

▲ How the horse's leg has changed over time

b **Explain what has happened to the other three fingers as the horse has evolved.**

Peppered moth

Although **evolution** by natural selection usually takes place over millions of years, it is possible to see natural selection taking place over a much shorter timescale.

During the industrial revolution, heavy pollution covered the trunks of trees and bushes with black soot. Before the pollution, the grey speckled 'peppered moth' had excellent camouflage on the bark of the trees. It was very difficult for birds to spot the moths and eat them.

Can you see the peppered moth in the photograph on the right?

When the trees became covered in soot, the moths became much more visible and were eaten by the birds.

Fortunately, owing to natural variation, some of the moths were slightly darker in colour. These moths had better camouflage against the dark trunks and were not eaten by the birds. These moths survived, and when they bred, they passed on the genes for darker wings to their offspring. Within a few years all the moths were dark and camouflaged.

Since the clean air act, the soot pollution has disappeared and the moth has evolved back into the grey speckled variety.

▲ This peppered moth will not survive as it is easily spotted by predators

 c **Suggest why the moths have evolved back into the grey speckled variety.**

Bugs and rats

Superbugs are resistant to nearly all of our antibiotic drugs. When antibiotics are used to treat disease, some of the bacteria may be slightly more resistant to the antibiotic than others. If these bacteria are allowed to survive, their resistance is passed on as the bacteria multiply. Soon all the bacteria are slightly resistant to the antibiotic. Because of variation, some of these bacteria have even more resistance. Within a few hundred generations, all the bacteria will be completely resistant to the antibiotic.

▲ This peppered moth is camouflaged

 d **Suggest why doctors often prescribe two different antibiotics for very serious diseases.**

Evolution and natural selection has also resulted in most rats now being resistant to the rat poison called warfarin.

Evolution of a new species

Sometimes a species may become separated into two different breeding populations. When this happens the two populations evolve independently. Sometimes they evolve so much that the differences prevent them from breeding with each other once the two populations come back together. When this happens, two new species have evolved.

 e **Suggest what physical barrier could separate a species into two different breeding populations.**

f **Suggest what evolved differences could prevent these two populations from breeding together.**

Keywords

Darwin • evolution • fossil • natural selection

▲ *Jean Baptiste de Lamarck*

Different theories

In 1800, Lamarck thought that evolution occurred because characteristics that were acquired during an organism's lifetime were passed on to its offspring. If this were true, suntans and tattoos would be passed on to our children. Lamarck's theory is now discredited, but he was the first person to notice that as the environment changed so did the organisms that inhabited it. This idea enabled Charles Darwin to realise that evolution occurred because of natural selection.

The mistake that Lamarck made was to think that individual organisms controlled the characteristics that were acquired, and that these were inherited.

Like many new ideas, Darwin's theory wasn't accepted by most people. The church refused to believe that humans could be related to apes. They misunderstood Darwin and thought he said we had evolved directly from apes, rather than having a common ancestor. Scientists now think that apes and humans are the result of millions of years of evolution, and that we both evolved from a common ancestor that lived many years ago. The diagram on page 45 shows this. Even today, in the face of all the evidence, some people still believe that evolution had no part to play in the diversity of organisms that live on Earth.

It is important that when we are faced with lots of different theories we look at the evidence before we come to a conclusion. Darwin knew that the study of science was about gathering data and evidence. Some people simply accept ideas because someone else told them it is true, but scientific thought must change in the face of new evidence.

In science, some theories are disproved as more evidence is gathered that contradicts the theory. Other theories, like the theory of evolution, become stronger as the evidence gathered supports it. It is important that as we study different ideas and theories in science we have an open mind and look for evidence.

Questions

1 Explain why Lamarck should be credited with increasing our understanding of evolution.

2 What mistake did Lamarck make with his theory?

3 Explain why Darwin's theory of natural selection was rejected by many people.

4 Suggest why Darwin's theory about the common ancestor for apes and humans will always remain a theory and may never be proved.

Pollution problems

In this item you will find out

- what effect the human population increase is having on the environment

- how different species can be used to monitor the level of pollution

As the human population increases, more and more of the Earth's limited **resources** of fossil fuels and minerals are being used up. Once used up, they are gone forever. Some scientists estimate that within the next ten years, half of all the Earth's crude oil will have been used up. As the population grows and more resources are consumed, **pollution** also increases.

In parts of the world the human population is growing **exponentially**. This means the population in those places doubles every 53 years.

If the human population continues to grow at this rate then there will come a point when the Earth will not be able to sustain all the people. Some countries, such as China, have already taken steps to slow down population growth.

 a If the human population is 6 676 871 265, calculate what the population will be after 24 hours.

Amazing fact

The human population gets bigger by more than two people every second.

◄ *The air over Los Angeles is quite polluted*

Who is doing the polluting?

Developed countries with smaller populations, consume more resources and pollute more than less developed countries with larger populations.

Place	Population (millions)	Carbon dioxide produced (billions of tonnes)
Africa	732	0.4
USA	265	4.9

b **Suggest why the USA has fewer people, but produces more carbon dioxide than Africa.**

Pollution has serious consequences for the environment.

Global warming

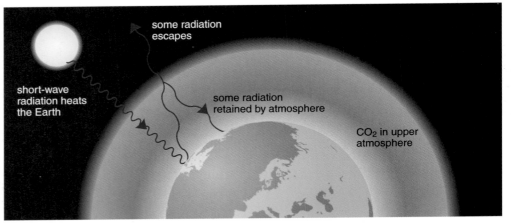

some radiation escapes

short-wave radiation heats the Earth

some radiation retained by atmosphere

CO_2 in upper atmosphere

▲ *The greenhouse effect*

Burning fossil fuel releases carbon dioxide into the atmosphere. We are now burning so much fuel that the level of carbon dioxide in the atmosphere is rising. Energy from the Sun hits the Earth's surface and causes warming. Normally this heat is radiated back into space. Carbon dioxide traps some of this heat energy in the atmosphere, just like the glass in a greenhouse traps heat. This causes the temperature of the Earth to rise. It is called the greenhouse effect.

Ozone depletion

The **ozone** layer is a layer of ozone gas that is found in the upper atmosphere. Ozone is normally harmful to humans, but in the upper atmosphere it absorbs most of the ultraviolet light from the Sun. Without the ozone layer, sunlight would contain so much ultraviolet light that it would not be possible for life to exist on the surface of the Earth. Unfortunately, many of the pollutants that we have produced, such as CFCs, are damaging the ozone layer. In recent years, a hole in the ozone layer has appeared over the Antarctic.

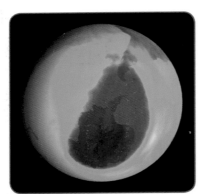

▲ *The hole in the ozone layer*

c **Explain why the ozone layer is so important to life on Earth.**

Acid rain

Most fossil fuels contain small amounts of sulfur. When the fuel is burned the sulfur is released into the atmosphere as sulfur dioxide. Sulfur dioxide dissolves in rain to form **acid rain**. The acid rain kills fish (as it turns rivers and lakes into dilute acid) and it kills trees. It also reacts with limestone buildings and dissolves them away.

Measuring pollution

Pollution can be measured using biological **indicator species**. Some freshwater organisms can only survive in clean water with lots of oxygen. If they are present in the river or stream we know the water is clean and pollution free. Other organisms can survive in polluted water with very little oxygen. If they are present in the river we know that it is polluted.

Organisms found in polluted water include the blood worm and water louse. Organisms found in very polluted water include the rat tailed maggot and sludge worm.

▲ This was caused by acid rain

▲ Blood worm

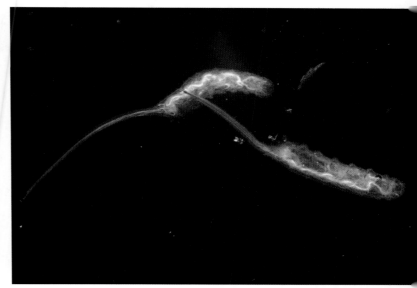

▲ Rat-tailed maggot

Air pollution can be measured using lichens.

Some other lichens are much less tolerant of pollution – if you find these lichens growing you know that there is very little pollution and that the air is very clean.

◀ This lichen can withstand moderate levels of pollutants such as sulfur dioxide. See if you can find it on buildings around your school.

Lichens as indicators

▲ Usnea

Different lichens can tolerate different levels of sulfur dioxide pollution. Sulfur dioxide destroys the chlorophyll that plants use for photosynthesis.

The table shows how much sulfur dioxide can be tolerated by three different lichens in $\mu g/m^3$.

Lichen	Level of sulfur dioxide tolerated ($\mu g/m^3$)
No lichens	175
Lecanora conizaeoides	125
Parmelia caperata	50
Usnea	0

▲ Parmelia caperata

Look at the map. It shows sulfur dioxide pollution in parts per billion.

The table of data and the map do not use the same units of measurement.

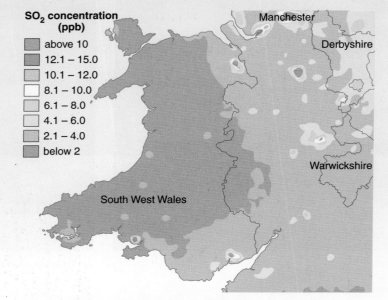

SO₂ concentration (ppb)

- above 10
- 12.1 – 15.0
- 10.1 – 12.0
- 8.1 – 10.0
- 6.1 – 8.0
- 4.1 – 6.0
- 2.1 – 4.0
- below 2

Manchester
Derbyshire
Warwickshire
South West Wales

▲ Map of Wales and Midlands showing sulfur dioxide pollution

▲ Lecanora conizaeoides

Questions

1 Suggest where the lichen *Usnea* may be found.

2 Suggest where the lichen *Lecanora conizaeoides* may be found.

3 The wind mostly blows from the south west. Suggest why there is less sulfur dioxide pollution in south west Wales.

4 Suggest why most lichens cannot tolerate high levels of sulfur dioxide pollution.

5 Suggest why there are high levels of sulfur dioxide in Manchester.

Extinction is forever

In this item you will find out

- why animals and plants become extinct

- how endangered species can be protected

- about sustainable development

▲ Leatherback turtle

When species die out, they become **extinct**. Some organisms become extinct naturally. This can happen when their environment changes owing to natural causes, such as **climate change**, or by competition with other more successful species. A species that is in danger of becoming extinct is **endangered**.

▼ Desmoulins whorl snail

a Describe what is meant by the phrase 'extinction is forever'.

Some animals may become extinct in the next few years because of the effect that humans are having on the environment.

Destroying habitats

In 1995, a new bypass was planned around the town of Newbury. The habitat of an endangered rare snail needed to be destroyed to build the road. Protestors tried to stop the road from being built, but the Newbury bypass was built and opened in 1998.

Fortunately some of the snails were saved and moved to another site.

Hunting

The Northern Right Whale is on the verge of extinction. During the 1800s, whaling ships reduced the population from thousands to just a few hundred. They were called Right Whales because they were the right whales to kill by the whaling ships.

Polluting

Leatherback turtles have existed for over 100 million years. They are likely to be extinct within the next ten years. One of the greatest threats is the turtles eating ocean pollutants (such as plastic bottles) that are thrown into the sea.

▲ Northern Right Whale

▲ Site of special scientific interest

How can humans help?

1 Protecting habitats

Habitats that contain rare or endangered species can be protected. They can be labelled sites of special scientific interest (SSSIs). This should prevent anyone developing the land and destroying the habitat.

2 Legal protection

Some species are given legal protection so that they cannot be hunted or killed. This law applies to many wild bird species.

3 Education programmes

People can be educated about how fragile and important our environment is. In this way they learn how to appreciate and respect it.

4 Captive breeding programmes

Some zoos have breeding programmes to breed rare species and release them back into the environment. Examples include birds of prey such as the red kite and the osprey.

5 Creating artificial ecosystems

Artificial ecosystems can be created for endangered species to live in. Even having your own wildlife pond can make a difference.

 Suggest another artificial ecosystem that could be created to help protect an endangered species.

▲ The red kite is a protected species

Conservation programmes make sense

Conserving the environment and all the different species within it is a very sensible thing to do. The more species that exist in a habitat, the more likely the habitat is to survive. It enables the habitat to withstand small changes.

This is because there are more food chains within a complex food web and therefore more opportunities for organisms to find food.

Humans are part of the environment. When we conserve it, it means that we protect our food supply as well.

We also obtain many of our medicines from plants. There are still thousands of undiscovered plants that could potentially provide us with new life-saving drugs.

People have also become very proud of some species, such as the oak trees of England or the golden eagles in Scotland. These plants and animals form part of our cultural inheritance and we should try to conserve them.

Examiner's tip

Make sure you know the difference between sustainable and non-sustainable development.

Sustainable development

One way that the environment can be protected is by **sustainable development**. This means that whatever is removed from the environment must be replaced.

Woodland that is cut down can be sustained by replanting with young trees. The young trees grow and sustain the woodland. In the future, these trees will also be harvested and replaced.

Many woodlands now have visitor centres that are used by schools to teach about conservation, including rare and endangered species.

Fish stocks in the North Sea are being seriously over-fished. The numbers of fish, such as cod, are falling dramatically. When the fish population drops below a critical level it may not recover.

Fishermen have been given quotas to limit the number of fish that they can remove.

This should then allow the fish to breed and the fish population to recover.

Many different countries fish in the North Sea. For sustainability to work, careful planning, cooperation and agreement are required between all the different countries.

C **At low tide, cockle pickers collect cockles from the sand and mud flats of Morecambe Bay. Suggest how the cockles could be conserved by sustainable development.**

Sustained woodland

Keywords

climate change • endangered • extinct • sustainable development

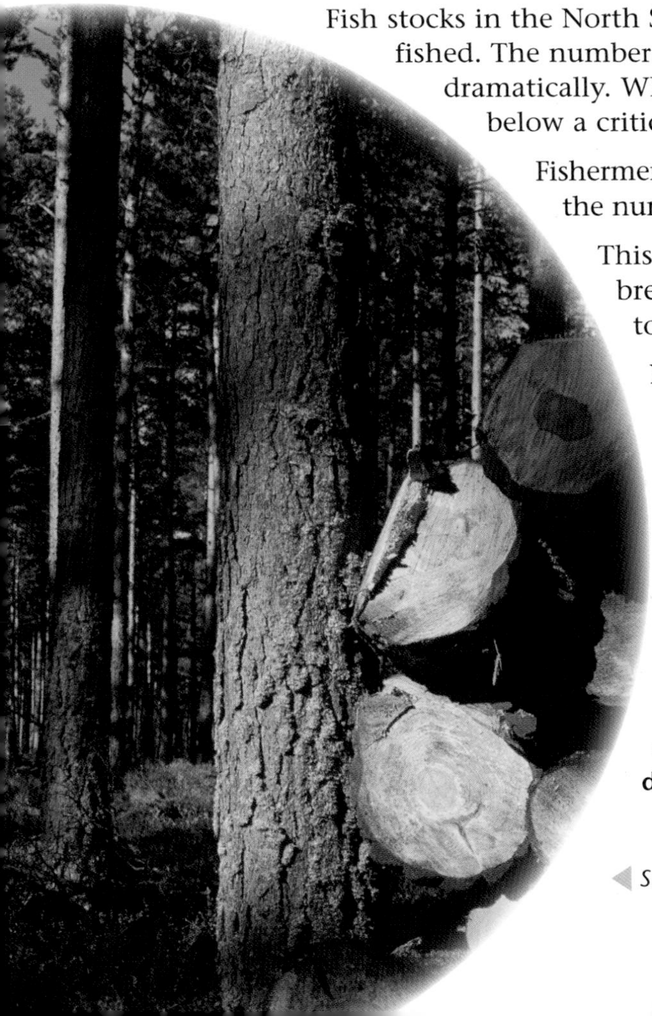

▲ Whales in the wild

Whale watching

Whales and dolphins are mammals. They are more closely related to humans than other marine animals. There is still much that we do not understand about whales.

They can dive and survive at great depths and are able to communicate over long distances. They also migrate from one part of the world to another to follow food supplies and mate. Marine biologists are still trying to discover how whales can do all of these things.

Whales are hunted in some parts of the world. They are valued for food and the oil that they contain. Chemicals from their body are also used for manufacturing some cosmetics. Some whales are killed for research.

Because whales live far from land it is not easy to get international agreements to stop whaling. It is even harder to police the laws and ensure that some whalers do not break these international agreements.

Humans are fascinated with whales and whale watching is a valuable commercial tourist industry.

Some whales and dolphins are kept in captivity. They are used for both entertainment and research. Many are bred in captivity and are not used to swimming in the open sea.

▲ Killer whale in captivity

Questions

1 Suggest why whale watching is such a popular tourist activity.

2 State two reasons why whales are still hunted by whaling boats.

3 State two things that we still do not fully understand about whales.

4 State two reasons why whales are kept in captivity.

5 Suggest why protecting whales is such a difficult thing to do.

6 Although whales and dolphins seem happy to perform for the public, explain whether you think they should be used in this way.

B2a

1 Describe the difference between an ecosystem and a population. [2]

2 There are many different types of ecosystem.
   a Name an ecosystem that has not yet been fully explored. [1]
   b Explain why it has not been fully explored. [1]

3 Use the key to identify the following organisms.

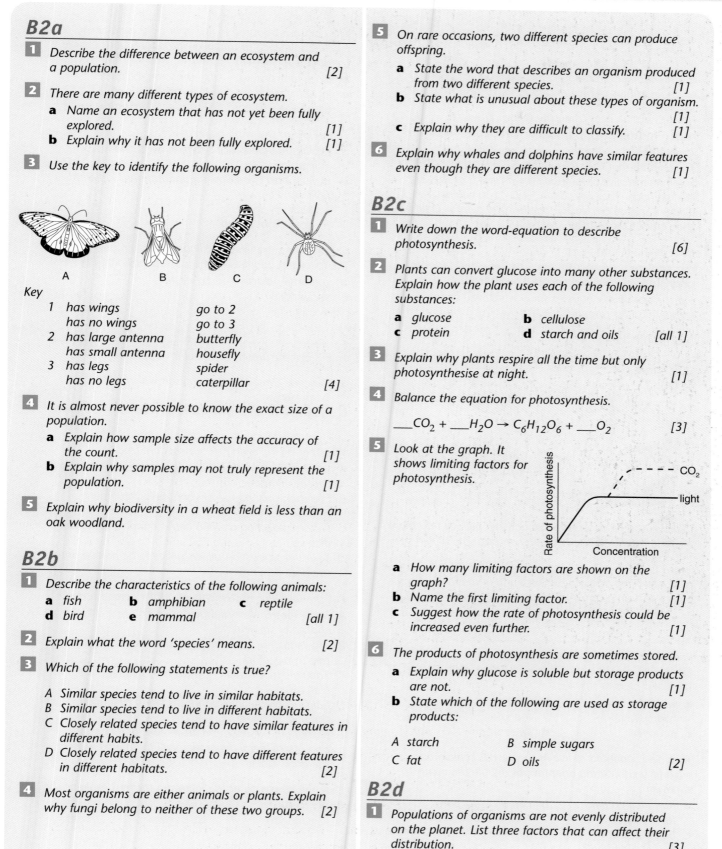

A   B   C   D

Key
   1 has wings   go to 2
      has no wings   go to 3
   2 has large antenna   butterfly
      has small antenna   housefly
   3 has legs   spider
      has no legs   caterpillar   [4]

4 It is almost never possible to know the exact size of a population.
   a Explain how sample size affects the accuracy of the count. [1]
   b Explain why samples may not truly represent the population. [1]

5 Explain why biodiversity in a wheat field is less than an oak woodland.

B2b

1 Describe the characteristics of the following animals:
   a fish   b amphibian   c reptile
   d bird   e mammal   [all 1]

2 Explain what the word 'species' means. [2]

3 Which of the following statements is true?

   A Similar species tend to live in similar habitats.
   B Similar species tend to live in different habitats.
   C Closely related species tend to have similar features in different habits.
   D Closely related species tend to have different features in different habitats. [2]

4 Most organisms are either animals or plants. Explain why fungi belong to neither of these two groups. [2]

5 On rare occasions, two different species can produce offspring.
   a State the word that describes an organism produced from two different species. [1]
   b State what is unusual about these types of organism. [1]
   c Explain why they are difficult to classify. [1]

6 Explain why whales and dolphins have similar features even though they are different species. [1]

B2c

1 Write down the word-equation to describe photosynthesis. [6]

2 Plants can convert glucose into many other substances. Explain how the plant uses each of the following substances:
   a glucose   b cellulose
   c protein   d starch and oils   [all 1]

3 Explain why plants respire all the time but only photosynthesise at night. [1]

4 Balance the equation for photosynthesis.

   ___CO_2 + ___H_2O → $C_6H_{12}O_6$ + ___O_2 [3]

5 Look at the graph. It shows limiting factors for photosynthesis.

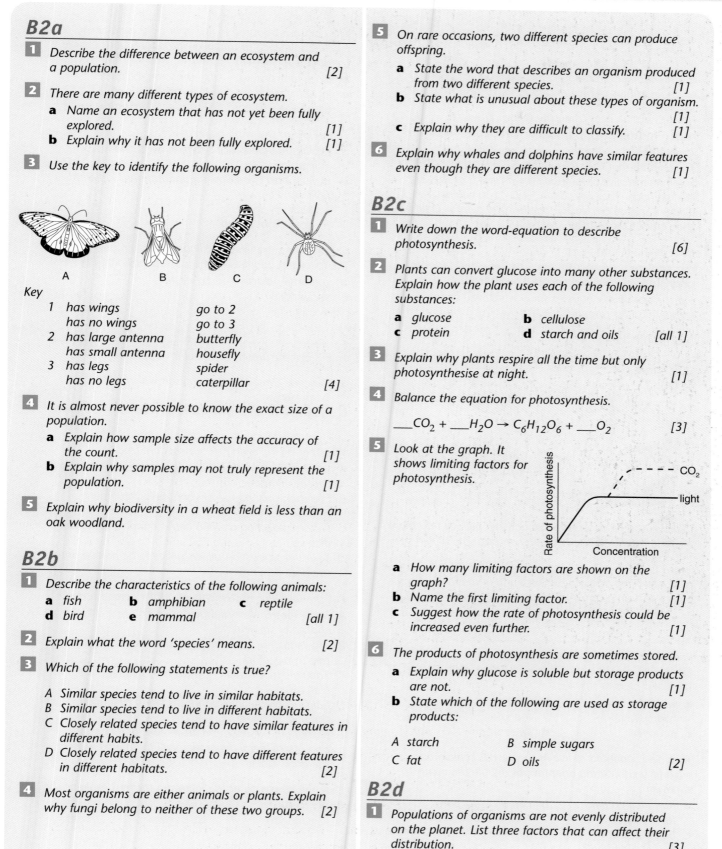

   a How many limiting factors are shown on the graph? [1]
   b Name the first limiting factor. [1]
   c Suggest how the rate of photosynthesis could be increased even further. [1]

6 The products of photosynthesis are sometimes stored.
   a Explain why glucose is soluble but storage products are not. [1]
   b State which of the following are used as storage products:

   A starch   B simple sugars
   C fat   D oils   [2]

B2d

1 Populations of organisms are not evenly distributed on the planet. List three factors that can affect their distribution. [3]

2 In a predator-prey relationship, describe what will happen when:

    a the number of predators increases   **[1]**
    b the numbers of prey decreases   **[1]**

3 Finish the sentences using the following words.

host   mutualism   parasites   predators   prey

Animals that feed on other animals are called __(1)__. The animals that they feed upon are called the __(2)__. Some animals live on or inside the animal that they are feeding upon. They are called __(3)__. Sometimes two different species live closely together and depend upon each other. This is called __(4)__.   **[4]**

4 Look at the following list of animals:

**red squirrel   hawk   falcon   red spotted ladybird
grey squirrel   yellow spotted ladybird**

Put the animals in pairs to show which will be in closest competition for resources.   **[3]**

5 Look at the graph.

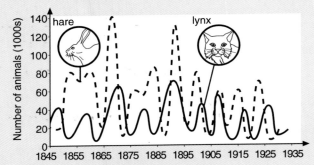

It shows the predator-prey relationship between the hare and the lynx. Describe two different patterns shown by the graph.   **[2]**

6 Nitrogen-fixing bacteria can be found in the roots of some plants. It is an example of mutualism. Explain why.   **[2]**

B2e

1 Polar bears are adapted to live in cold arctic conditions. Explain how each of the following adaptations enable the polar bear to survive:

    a white fur for   **b** layer of fat under
       camouflage   the skin
    c large feet   **d** large size   **[all 1]**

2 A camel is adapted to live in the desert. Explain how each of the following adaptations enable the camel to survive:

    a hump containing   **b** can allow its body
       fat   temperature to rise
    c bushy eyebrows   **d** large feet
       and hairy nostrils   **[all 1]**

3 The cactus is adapted to live in hot dry conditions. Explain how each of the following adaptations enable the cactus to survive:

    a large root system   **b** thick tough cuticle
    c lots of storage   **d** large volume to
       tissue   surface area ratio   **[all 1]**

4 Some flowers are wind pollinated.

    a Explain the difference between wind and insect
       pollination.   **[1]**
    b Explain how the following adaptations help wind
       pollinated flowers:
       i feathery stigmas
       ii small light pollen   **[2]**

5 Some flowers are insect pollinated. Explain how each of the following adaptations help insect pollinated flowers:

    a colourful petals   **b** nectar
    c sticky pollen   **[all 1]**

B2f

1 Fossils are the preserved remains of dead organisms.

    a Describe three different ways in which fossils can
       be preserved.   **[3]**
    b Give two reasons why the fossil record is incomplete.
             [2]

2 Which of the following statements are true?
    A When the environment changes some organism may
      become extinct.
    B Evolution only happens when the environment
      changes.
    C Natural selection is when better adapted organisms
      die.
    D Sexual reproduction allows genes to be passed onto
      the next generation.   **[3]**

3 Put the following statements into their correct order.
    A Some moths were slightly darker and had a better
      chance of survival.
    B The industrial revolution produced soot that turned
      tree trunks black.
    C The pale moth could now be seen by predators and
      was eaten.
    D Pale-coloured peppered moths were camouflaged on
      the bark of trees.
    E Sexual reproduction produced even darker moths
      and after several generations all the moths were of
      the dark variety.   **[5]**

4 Describe how each of the following theories explains the fossil record:

    a Creationism   **b** Darwinism   **[both 1]**

5 Which of the following statements about natural selection are true?

A Natural selection can only occur if all the organisms have the same genes.

B Sexual reproduction produces variation,

C In a changing environment, organisms compete for limited resources.

D Owing to variation, some organisms are better adapted than others to the new environment.

E Species that are unable to compete may become extinct. [5]

6 Some characteristics are acquired during the life of the organism.

a Explain whether these characteristics can be inherited. [1]

b Explain the difference between Lamarck's and Darwin's theory of evolution. [1]

B2g

1 Link the following words to the correct statements:

carbon dioxide   CFCs   sulfur dioxide

A acid rain

B global warming

C destruction of ozone layer [3]

2 Look at the graph. Describe the relationship (pattern) between population growth, use of resources and pollution.

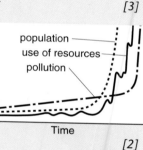

Time

[2]

3 The following organisms were found in three different streams:

Stream A – caddis fly and may fly larvae
Stream B – blood worm and rat tailed maggot
Stream C – stone fly larvae and water scorpion.
Which stream is the most polluted?
Explain your answer. [2]

4 Look at the following data.

Country	Population (millions)	Carbon dioxide produced (billions of tonnes)
Africa	732	0.4
America	120–170	4.9

a Explain the difference in figures between Africa and America. [1]

b Populations in parts of the world are increasing exponentially. Explain the consequences of this increase. [1]

5 Look at the graph. It shows sulfur levels in a lichen growing up to 120 km from an isolated city in America.

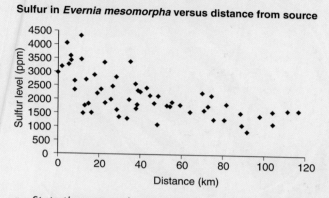

Sulfur in *Evernia mesomorpha* versus distance from source

a State the pattern between sulfur levels in the lichen and distance from the city. [1]

b Suggest why different lichens at the same distance from the city have different sulfur levels. [1]

c Suggest a possible source of the sulfur. [1]

B2h

1 Match the following reasons for extinction with the correct description:

**climate change   habitat destruction   hunting
pollution   competition**

A fishing for North Sea cod

B two different organisms needing the same source of food

C global warming

D cutting down the rain forests

E the release of sulfur dioxide from fossil fuels

2 Humans can help endangered species. Which of the following are methods that will be successful?

A legal protection of their habitat

B increase fishing and hunting licences

C captive breeding programmes

D creating artificial ecosystems

E building new roads and towns [2]

3 Conservation is important. Give two reasons why conservation is important. Explain your answer. [4]

4 Explain why it is easier to conserve red squirrels than whales. Explain your answer. [2]

5 Describe how harvesting wood from a forest can be made into a sustainable development. [1]

- Every day, newspapers contain articles about new breakthroughs in genetics. They include stories of how scientists are going to produce new organisms by manipulating genes.

- All this is possible due to the discovery of the structure of DNA and how it acts as the genetic code.

- Life on our planet ranges from the smallest single-celled organisms to massive organisms made of millions of cells. Although there are advantages to being larger, these cells need organising and transport systems are needed to move substances around them.

I read in the papers that scientists can now make identical copies of any animal by cloning them.

I'm not sure that cloning would make identical copies of a person. Surely how we are brought up is just as important as our genes.

I don't agree with cloning people but it must be useful to be able to clone animals.

What you need to know

- About the reproductive system and the breathing system.

- What plants need so that they can make their own food by photosynthesis.

- The importance of proteins in the body.

- The structure of plant and animal cells.

Building blocks for organisms

In this item you will find out

● where respiration takes place in cells

● about DNA and DNA 'fingerprints'

● the function of enzymes and how they work

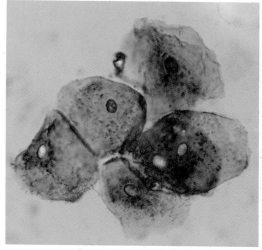

▲ Human cheek cells

▼ Canadian pondweed cells

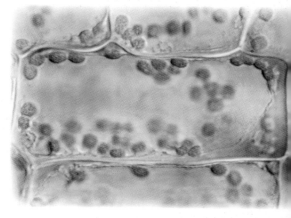

All living organisms are made up of cells. Using the naked eye we can see detail down to about the size of 0.1 mm. Cells are smaller than this so it is necessary to use microscopes to study them.

The first person to see cells was a British scientist called Robert Hooke in 1665. He looked at cork under the microscope and saw little boxes. He called them cells because he thought that they looked like the rooms where monks slept.

As microscopes improved, more and more detail could be seen in cells. You have probably used a light microscope to look at cells and seen images like the cells in the photographs.

a What structures can you see in these cells?

By using a light microscope, we can see objects as small as 0.002 mm clearly. About 50 years ago a new type of microscope was developed called the electron microscope. This could show even more detail in cells – down to about 0.000 002 mm.

Although this is small enough to see much of the detail in cells, this is still not small enough to see individual atoms. Scientists had to find other ways of investigating the molecules that are found in cells. One of these molecules is DNA.

Two scientists, James Watson and Francis Crick, used images obtained by firing X-rays at a crystal of DNA. They used these data to work out the structure of the DNA molecule in 1953.

Amazing fact

DNA is made of molecules called bases which are about 0.000 000 34 mm wide.

▲ Watson and Crick

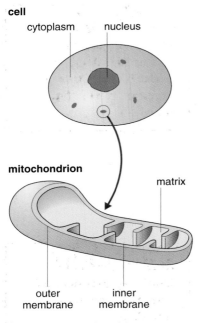

cell
cytoplasm   nucleus

mitochondrion

matrix

outer
membrane

inner
membrane

▲ *Inside a mitochondrion*

Inside cells

It is now possible to look more closely at the cytoplasm of animal cells. Using the electron microscope, the detail of structures such as mitochondria can be studied.

Mitochondria contain enzymes that carry out the final stages of respiration and this provides all the energy for life processes.

b Why do you think the inner membrane of the mitochondrion is so folded?

Making proteins

We have known for a long time that the nucleus of the cell is the control centre. It regulates the activities of the cell by controlling which proteins the cell makes (protein synthesis). We now know that DNA is present in the nucleus of every cell where the code is kept. So, it is likely that DNA contains the genetic code.

c Suggest what the functions of these proteins might be?

Each different protein is made up of a specific chain of amino acids. This determines the shape of the protein and how it works. So, DNA must code for the order of the amino acids.

We get the amino acids that are used by the cell to make proteins from our food. Some foods do not contain a very wide variety of amino acids. Fortunately, some amino acids can be converted into others in the liver (transamination) but essential amino acids must be obtained from our diet.

Structure of DNA

Watson and Crick found that DNA was made of two strands of organic bases twisted up to make a spiral or a **double helix**. There are four different bases, known by the letters A, T, G and C. Pairs of these bases form cross links that hold the two chains together. Base A always pairs up with T and base G always pairs up with C. This is called **complementary base pairing**.

The order of the bases codes for the order of amino acids in each protein. Each amino acid is coded for by a sequence of three bases. This is shown in the diagram below.

▲ *The structure of DNA*

triplet sequence

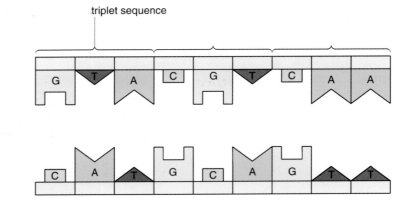

▲ *Strand of DNA showing detail of bases*

The whole base sequence that codes for one protein is called a gene. Each protein has its own number and sequence of amino acids, which means that each protein has a different shape and function.

Before cells divide, the DNA copies itself (DNA replication). The two strands of the DNA molecule can unzip and come apart to form single strands. As each part of the strand becomes unzipped, new bases can move in to form a double strand again by complementary base pairing. At the end of the process there are two complete double strands of DNA. Because the bases can only fit together in certain ways, both double strands of DNA are exactly the same. When the cell divides each copy of the cell has the same DNA.

Enzymes

Enzymes are one of the most important types of protein produced by all cells. Enzymes are biological catalysts. This means that they speed up the rate of chemical reactions in living cells, such as respiration, photosynthesis and protein synthesis.

All enzymes also share the following characteristics:

- they have an optimum temperature and pH where they work best
- each enzyme will only work on a particular reaction.

These properties of enzymes can be explained by looking at how they work. The chemical that is reacting (the **substrate**) fits into an area on the enzyme called the **active site**. The reaction then takes place. The enzyme and the substrate are like a lock and key. This explains why enzymes are specific to a substrate and a reaction.

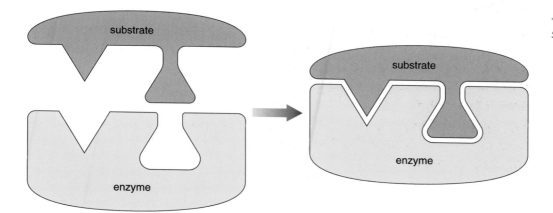

Most enzymes work best at body temperature and at the pH of the organ in which they work. High temperatures and extremes of pH will change the shape of the enzyme's active site irreversibly. The enzyme is said to be **denatured**. The lock and key theory explains why this will stop the enzyme working.

Amazing fact

Even though a cell is less than 0.03 mm wide the DNA that it contains is over 1 m long!

Examiner's tip

Don't say that heat *kills* enzymes. Enzymes are not alive.

◀ *An enzyme fits into a specific substrate*

Keywords

active site • complementary base pairing • denatured • double helix • mitochondrion • substrate

Making a DNA fingerprint

Dr Rachel Wong works in a laboratory that deals with DNA testing. Although most of our DNA controls protein synthesis, some does not. This non-coding DNA is unique to each person and so can be used to identify an individual. It is called a DNA 'fingerprint'.

'Whenever I create a DNA fingerprint I go through several stages,' she says. 'The DNA is isolated and extracted from blood cells. Then it is cut into fragments using a special enzyme.

The DNA fragments are separated into bands during a process called electrophoresis. The bands are now embedded in gel but we can't see them. This band pattern is then transferred to a nylon membrane.

A radioactive DNA probe is added which binds to the bands of DNA. The extra radioactive DNA probe is washed off.

When the nylon membrane is X-rayed the radioactive DNA probe shows the pattern of the DNA fingerprint. This fingerprint can then be compared with another fingerprint.'

Rachel has just finished preparing the DNA fingerprints of a man, a woman and two children.

1. Blood sample   2. DNA extracted from blood cells and cut into fragments by a restriction enzyme

3. DNA fragments are separated into bands during electrophoresis on an agarose gel

5. The radioactive DNA probe is prepared

7. At this stage the radioactive probe is bound to the DNA pattern on the membrane. An X-ray film placed next to the membrane reveals the pattern of bands known as a DNA fingerprint

6. DNA probe binds to specific DNA sequences on the membrane

4. DNA band pattern in gel is transferred to a nylon membrane by a technique known as Southern blotting

child 1   mother   child 2   possible father

Questions

1 Why do you think a radioactive DNA probe is added before the DNA band pattern is exposed to an X-ray film?

2 Which of the children could be the offspring of the woman and the man?

3 What would the genetic fingerprint of identical twins look like?

4 Describe one other possible use of genetic fingerprinting apart from proving relationships between people.

Spreading far and wide

In this item you will find out

- how diffusion happens
- the importance of diffusion in animals
- how diffusion occurs in plant leaves

One morning in May during the 1870s, the French scientist Jean-Henri Fabre was pleased when a female great peacock moth emerged from a cocoon on a table in his laboratory-study.

He put her under a wire cage and left her to spread her wings to dry. That evening Fabre was amazed as dozens of male great peacock moths floated in through the open doors and windows of his house. Over the following week Fabre caught more than 150 male moths. No matter where in the house he moved the female, the male moths headed directly for her. 'What was attracting them?' he wondered.

Over the next several years Fabre carried out many experiments to learn the moths' secret. Eventually he concluded that, even though humans could not detect it, the female moth must release a smell that is very attractive to the opposite sex of her species.

a Most male moths have long feathery antennae. What do you think they are used for?

We now call the chemicals that the moth was producing pheromones. They are produced by many animals and can attract mates from many kilometres away. But how do these chemicals travel such long distances in the air?

The answer is diffusion. Diffusion allows many chemicals to spread through the air. Insects are attracted to flowers by scent diffusing into the air.

Diffusion is also important inside the bodies of animals and plants. Many substances rely on diffusion to enter and leave the body and its tissues.

▲ *A female great peacock moth*

▼ *This insect is attracted by the flower's scent*

What is diffusion?

Diffusion is described as the net movement of a substance from an area of high concentration to an area of low concentration. It works because individual particles are constantly moving about at random and so they tend to spread out.

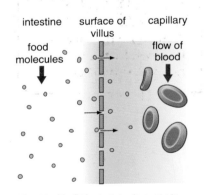

▲ What cells gain and lose

A plant or animal cell will lose waste products and gain useful substances by diffusion through the cell membrane. This is shown in the diagram on the left.

The rate of diffusion of a substance can be made faster by:
- decreasing the distance the substance has to diffuse
- increasing the difference in the concentration between the two areas (the **concentration gradient**)
- making the surface area through which the substance has to diffuse larger.

b How can the surface area of a cell membrane be made larger?

Diffusion in animals

Substances can pass in and out of cells by diffusion and this allows animals to exchange substances with their surroundings.

In the small intestine, small digested food molecules are absorbed into the bloodstream by diffusion. The inside of the small intestine is long, permeable and covered with finger-like projections called **villi**. They are further covered by smaller projections called **microvilli**. Both the villi and the microvilli increase the surface area, speeding up absorption. The villi have a good blood supply as they contain lots of blood vessels. The food molecules can be taken away after being absorbed into the blood by diffusion.

▲ Food molecules are absorbed into the blood by diffusion

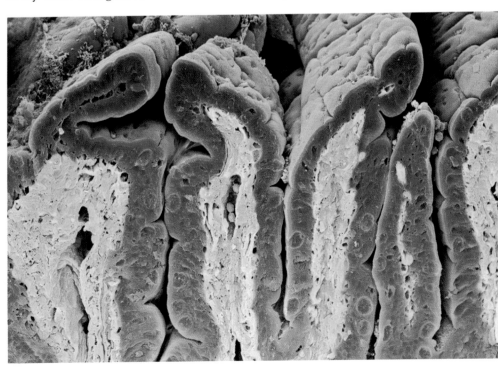

Villi inside the small intestine ▶

c Why do you think it is important to keep removing the food molecules that have entered the villi?

In the lungs, oxygen diffuses into the bloodstream and carbon dioxide diffuses out. This happens in small air sacs called **alveoli**. There are millions of these alveoli so that the total surface area is enormous. Like the villi, their walls are permeable and very thin (only one cell thick) so that the gases do not have far to diffuse. They also have a moist lining and a rich blood supply.

A **fetus** needs food and oxygen while it is growing – it gets them from its mother. In return, carbon dioxide and other waste products pass to the mother. Early in pregnancy the embryo grows a structure called the **placenta**. This is a disc of tissue that is attached to the wall of the uterus. All of these substances diffuse across the placenta. The placenta has projections called villi that increase the surface area for diffusion.

Diffusion is also important in the nervous system. One neurone is not directly connected to the next – there is a small gap. This is called a **synapse**. When a signal reaches the synapse, chemicals called **transmitter substances** are released and diffuse across the gap. This will set off a signal in the next neurone.

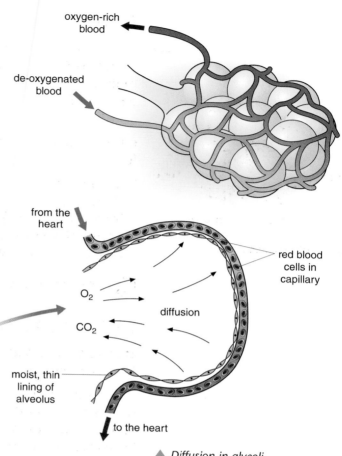
▲ *Diffusion in alveoli*

 d If all neurones in the body were directly connected, what would happen if one was stimulated?

Diffusion in plants

In plants, oxygen and carbon dioxide diffuse in and out through the leaves. Leaves are specially adapted to allow gases to diffuse in and out. They have thousands of microscopic pores called stomata on the underside, and the thin but flat shape of the leaves gives them a large surface area. Inside the leaf are air spaces that allow the gases to diffuse. All of these features increase the rate of diffusion of oxygen and carbon dioxide.

 e What processes produce **oxygen and carbon dioxide** in the plant?

 f How will the diffusion of these gases differ at night compared with during daylight?

As the stomata let gases diffuse in and out of the leaf, they will also allow water molecules to diffuse out. Plants lose large amounts of water from their leaves by this evaporation. This means they have to take up water from the soil in order to replace the water that is lost.

> **Examiner's tip**
>
> Remember the placenta is grown by the baby not by the mother.

> **Keywords**
>
> alveolus • concentration gradient • fetus • microvillus • placenta • synapse • transmitter substance • villus

Sugar and eggs

Angus is carrying out an experiment to show diffusion of water molecules across the membrane of an egg. He removes the shell of an egg with ethanoic acid, leaving the membrane of the egg plus the white and the yolk. An egg contains lots of water so it is an area of high concentration of water.

He places the egg in a sugar solution which is made of pure sugar with only a small amount of water in it. This means the sugar solution is an area of low concentration of water.

After the egg has been in the syrup for about an hour, Angus can see a layer round the egg in the syrup. The water molecules inside the egg have diffused through the membrane into the syrup and this layer is an area where the water concentration of the syrup has increased.

After 36 hours, most of the water in the egg has moved into the sugar solution. Only the yolk and the egg membrane are left. The water molecules have moved from the area of high concentration to the area of low concentration until the water concentration in the egg and in the syrup is almost equal.

▼ *Before*

sugar solution

white

yolk

▼ *After 36 hours*

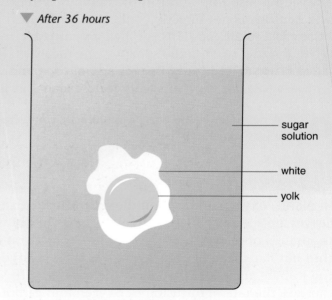

sugar solution

white

yolk

Questions

1 Why is the shell of the egg removed with ethanoic acid?

2 Why does a layer form round the egg after an hour?

3 After 36 hours the only parts of the egg left are the yolk and the membrane. What does this tell you about the egg white?

4 This experiment was carried out at room temperature. Suggest what would happen if the egg and the sugar solution were heated.

Blood is thicker than water

In this item you will find out

- how blood cells are adapted for the job they do

- how the heart and blood vessels move the blood

- about the problems of replacing hearts

For thousands of years people have realised the importance of blood to life. People used to believe that blood contained mystical powers and could not be divided into smaller parts.

We now know that blood is made up of different components and we understand many of the functions of the different parts. This means that we can predict how changes in the blood can affect us. Doctors can also make changes to the blood.

The blood contains three types of cells and they all carry out different jobs. The most numerous are the **red blood cells**. They carry oxygen around the body.

A red blood cell is shaped like a small disc with a dent in each side. This shape is called a biconcave disc. This is shown in the diagram on the right.

Many features of a red blood cell make it well adapted for the job of carrying oxygen around the body:
- they are small cells so they can fit through the narrowest blood vessels
- they contain a red protein called **haemoglobin** which carries the oxygen around the body
- their small size and biconcave shape gives them a large surface area/volume ratio so that they can lose or gain oxygen more quickly
- they don't have a nucleus so that more haemoglobin can fit in.

At the lungs the haemoglobin combines with oxygen. This forms **oxyhaemoglobin**. When the red blood cells reach body tissues the oxyhaemoglobin releases the oxygen.

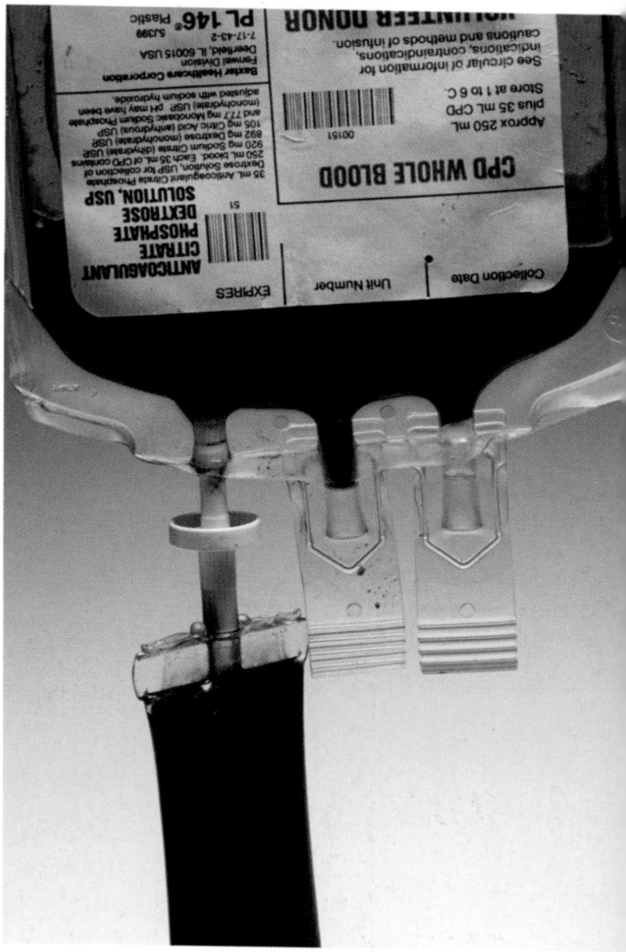

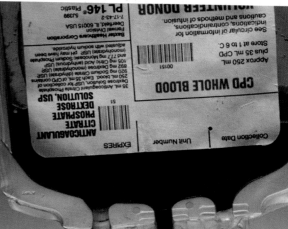

▲ Our understanding of blood means we can give people blood in emergencies

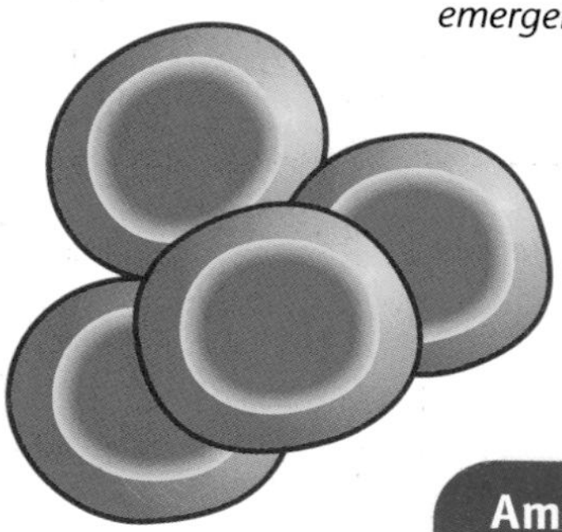

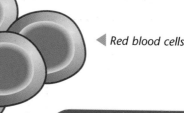

◀ Red blood cells

Amazing fact

Every cubic millimetre of blood contains about five million red blood cells.

83

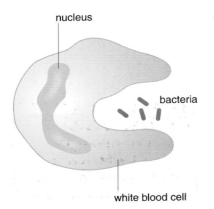

▲ *White blood cells attack bacteria*

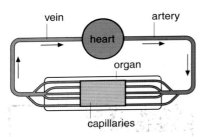

▲ *Different types of blood vessel*

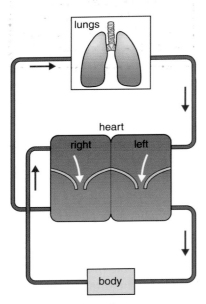

▲ *The double circulatory system*

White blood cells and plasma

The blood also contains **white blood cells**. They are twice as large as red blood cells and there is one white cell to every 500 red cells. They engulf and destroy disease organisms. To allow them to do this they can change shape so that they can squeeze out of the blood vessels and surround the invader.

All the cells are carried around the body by the liquid part of the blood called **plasma**. Plasma is mostly water but it also contains hormones, dissolved food, antibodies and waste products.

The pipework of the body

The blood is carried around the body in blood vessels. There are three types of blood vessels: **arteries**, **capillaries** and **veins**. Arteries carry blood away from the heart and veins carry blood back to the heart. Capillaries join arteries to veins and it is here that substances can be exchanged with the tissues.

The three types of blood vessels are each adapted for their particular function:
- arteries have a thick, muscular and elastic wall because the blood is under higher pressure than blood in the veins
- veins have a large lumen to reduce resistance to flow, and valves to prevent the blood flowing back because it is under low pressure
- capillaries are permeable so that substances can be exchanged with the tissues.

How the heart works

Mammals have **double circulation**. This means that the blood has to pass through the heart twice on each circuit of the body. Deoxygenated blood is pumped to the lungs and the oxygenated blood returns to the heart to be pumped to the body. The advantage of this system is that the pressure of the blood stays quite high and so it can flow faster around the body.

Because of the double circulation the heart is actually two pumps in one. The right side pumps the blood to the lungs. The left side pumps it to the rest of the body.

The heart is made up of four chambers. The top two chambers are called **atria** – they receive blood from veins. The bottom two chambers are **ventricles** – they pump the blood out into arteries. The top two chambers, the atria, fill up with blood returning in the **vena cava** and **pulmonary** veins. The two atria then contract together and pump the blood down into the ventricles. The two ventricles then contract, pumping blood out into the **aorta** and pulmonary arteries at high pressure.

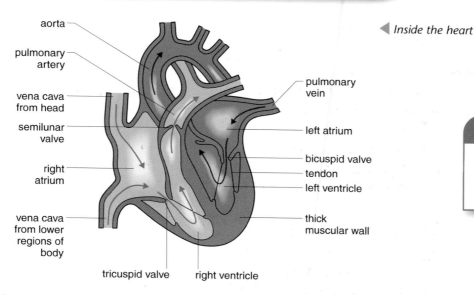

aorta

pulmonary artery

vena cava from head

semilunar valve

right atrium

vena cava from lower regions of body

tricuspid valve

right ventricle

pulmonary vein

left atrium

bicuspid valve

tendon

left ventricle

thick muscular wall

◄ *Inside the heart*

Examiner's tip

Remember the right side of the heart is on the left as you look at it.

The muscle wall of the left ventricle is always thicker than that of the right ventricle. This is because it has to pump blood all round the body compared with the short distance to the lungs.

In the heart are two sets of valves. The function of the valves is to prevent blood flowing backwards. In between the atria and the ventricles are the **bicuspid** and **tricuspid** valves. These valves stop blood flowing back into the atria when the ventricles contract. The pressure of blood closes the flaps of the valves and the tendons stop the flaps turning inside out. There are also **semilunar** valves between the ventricles and the arteries.

Too much fat?

Most people now realise that too much fat is bad for the heart. It is not all types of fat but two main types, saturated fats and **cholesterol**. Large amounts of these fats in the diet can cause cholesterol to build up in arteries. This happens over a long time and may form a blockage called a **plaque**. This may reduce or even stop blood flow and damage the heart. If the heart is badly damaged then the whole heart may need to be replaced.

Keywords

aorta • artery • atrium • bicuspid • capillary • cholesterol • double circulation • haemoglobin • oxyhaemoglobin • plaque • plasma • pulmonary • red blood cell • semilunar • tricuspid • vein • vena cava • ventricle • white blood cell

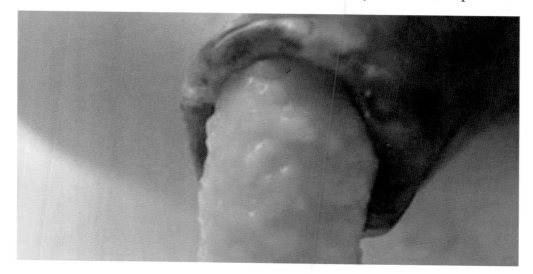

◄ *Cholesterol build-up in an artery*

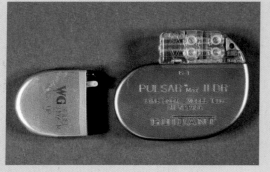

▲ A heart pacemaker

Replacing hearts

Derek Jones has been having problems with his heart for a couple of years. His doctor has sent him for some tests and the results have just come back.

'As we thought, your heart is beating irregularly. Also, your bicuspid valve is not working properly,' says Dr Galloway. 'We have two choices. We could fit you with an artificial pacemaker and an artificial valve. The pacemaker will regulate your heartbeat. Alternatively, we could give you a human heart from a donor.

'If we use a pacemaker and a valve, they are available now. However, there are problems with mechanical replacements. The pacemaker will need a battery to supply it with power which will need to be replaced regularly. Also, some people have bad reactions to the materials used to make the artificial parts. It is also difficult to get parts in the right sizes. It is harder for a pacemaker to change its rate during exercise than it is for a real heart.'

'What about a human heart?' asks Derek.

'Well, we can wait for a human donor heart to become available but there is a lack of donors. Once a heart is found we need to make sure that it is the correct size, age and tissue match for you. Even after successful surgery, you will need to take drugs to stop you rejecting your new heart.'

Questions

1 Why do you think the bicuspid valve not working properly in Derek's heart is a problem?

2 What does a pacemaker do?

3 What are the disadvantages of being fitted with a pacemaker?

4 At present many more people are given pacemakers than heart transplants. Why do you think this is the case?

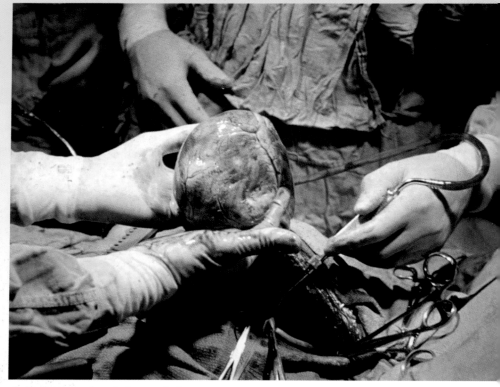

▲ Heart transplant being carried out

Cell multiplication

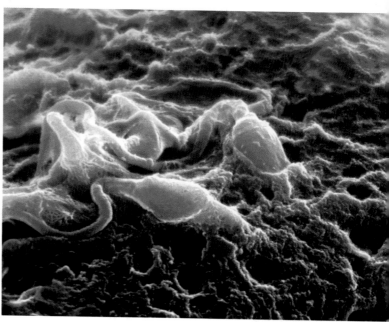

The number of cells in this fetus is increasing all the time

In this item you will find out

- how organisms produce new cells for growth

- the advantages to an organism of being multicellular

- how organisms produce sex cells for reproduction

We all know that every organism is made up of cells – in the case of humans, a very large number of cells. One major question that scientists have asked for hundreds of years is where do all these cells come from?

Although sperm were first seen under the microscope in 1670, it was over 200 years before their role in fertilising the egg was understood. For a long time scientists thought that the sperm carried all the information needed to make a baby.

It was over 50 years later that scientists saw eggs under the microscope.

a Why do you think scientists saw sperm before eggs even though eggs are much larger?

When sperm were first seen fertilising eggs, scientists realised that both the eggs and the sperm must carry important information. However, they did not know how.

When chromosomes were seen in the nucleus of a cell, scientists realised that they must carry the genetic information for the new organism. This work made scientists wonder how cells divide to make eggs and sperm, and how the joining of these two cells can produce an organism made of millions of cells. Once these questions were answered, scientists could use this information to manipulate reproduction in many different ways. Scientists are now wondering if they can use knowledge about cell division to find answers to other problems, such as how to stop people growing old.

A sperm fertilising an egg ▶

Cell division for growth

In order for an organism to increase in size or grow, its cells can become larger. However, there seems to be a limit to the size of cell a nucleus can control. Most growth involves cells dividing so that they can increase in number. The type of cell division that is used for growth is called **mitosis**.

In the body cells of mammals, the nuclei contain two copies of each chromosome. They are said to be **diploid**. In humans there are 23 pairs in each body cell making 46 chromosomes.

In mitosis the chromosomes are copied and the copies stay joined together. The chromosomes then line up down the middle of the cell. The copies of the chromosomes are then pulled apart to opposite ends or poles of the cell. When the cell divides into two, each new cell gets a copy of each pair of chromosomes, so they are genetically identical and still diploid.

Multi-cellular v unicellular

There are many advantages to being **multi-cellular**. It allows an organism to have different cells specialised to do different jobs. This is called **differentiation** and makes the cells more efficient. An organism can also become larger and more complex. This means that it can protect itself more easily. Lots of small cells also have a larger surface area than one large cell and so more materials can move in and out of the cells.

b Many factories have differentiation in the jobs that their workers are trained to do. What can happen to the factory if a particular worker is not able to work?

c What is a possible disadvantage of differentiation in the body?

Cell division for reproduction

Some organisms can reproduce by splitting into two, but this produces identical offspring. This is because their cells are all made by mitosis. Another type of reproduction is sexual reproduction. This involves sex cells (**gametes**) combining at **fertilisation** to form a **zygote**.

The sex cells, the sperm and eggs, have half the number of chromosomes of a normal body cell. In humans, the sperm and eggs each have 23 chromosomes instead of the usual 46. They are said to be **haploid**. This means that when the sperm and the egg join, the number of chromosomes becomes 46 again and the zygote has the diploid number of chromosomes. This is shown in the diagram on the right.

a cell has four chromosomes, two pairs

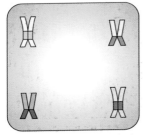

chromosomes are copied

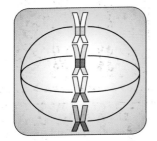

chromosomes form one line down the centre of the cell

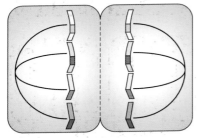

one copy of each chromosome moves to the opposite pole of the cell

▲ *Mitosis*

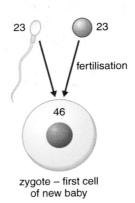

zygote – first cell of new baby

▲ *Gametes combine to form a zygote*

The type of cell division that makes gametes is called **meiosis**. This results in each gamete having a different combination of genes. Also, during fertilisation, any sperm can fertilise any egg. This means that the zygote can have any one of a number of possible combinations of genes. This explains why all organisms look different and how meiosis introduces variation into a species.

Before cell division each chromosome is copied. Pairs of chromosomes line up side by side. Each member of the pair is then pulled apart and they move to the opposite ends or poles of the cell. The cell then divides and then each chromosome divides and is separated. The cell then divides again to produce four cells. Each cell contains half the number of original chromosomes.

Sperm

Sperm cells have adaptations that allow them to carry out their function:
• they have lots of mitochondria to provide them with energy
• they have an **acrosome** which is a sac containing enzymes that enable them to digest the egg membrane and get into the egg.

 d What is the job of a sperm cell?

 e Why does it need lots of energy?

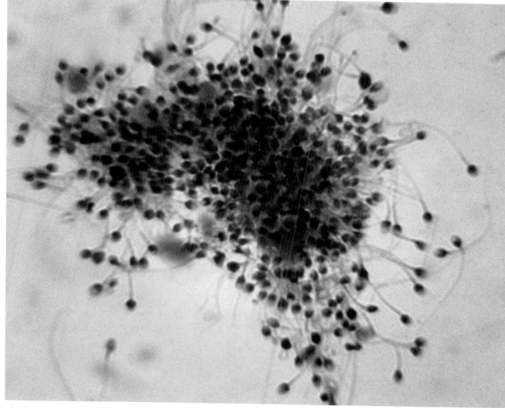

▲ *Sperm cells*

A cell has four chromosomes, two pairs

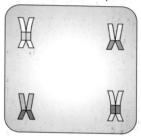

chromosome are copied

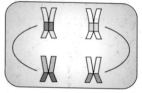

chromosome pairs line up side by side

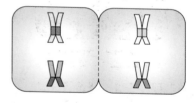

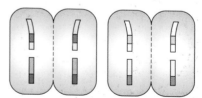

the copies split to produce four cells, each containing half the original number of chromosomes

▲ *Meiosis*

Keywords

acrosome • differentiation • diploid • fertilisation • gamete • haploid • meiosis • mitosis • multi-cellular • zygote

Can we live forever?

As we age, our bodies get worse at repairing damage to our tissues and replacing cells that have worn out. This is what getting old is all about. Even if doctors can stop us catching diseases, we cannot live forever because our bodies just wear out.

Scientists are now looking at why this is. Every time our cells divide, our genetic material or chromosomes have to copy themselves. It appears that our chromosomes have a 'protective cap' on their ends, just like shoelaces have plastic caps. These caps are called telomeres. Every time a cell divides these caps get shorter. When they reach a certain length the cell cannot divide any more. The person is getting old.

Scientists have now found that the cells that make the sex cells have an enzyme called telomerase. This repairs the telomere so that the sex cells have a full-length cap on their chromosomes.

The interesting question now is: 'Could all cells be made to produce this enzyme and make us live forever?' Some cells do have this ability, but the problem is they are cancer cells.

Questions

1 What is the name of the type of cell division that makes new cells for growth?

2 Why is it important that the chromosomes copy themselves exactly in this type of cell division?

3 A person receives a heart transplant from a much older person. Why is the discovery of telomeres rather worrying for this person?

4 Cancer cells have the enzyme telomerase. What can cancer cells do that normal body cells cannot do?

5 If we can only live for a certain maximum number of years, why do you think the average age that people live is increasing?

Growth spurts

In this item you will find out

- the differences between plant and animal cells

- how plants and animals grow

- how cells become specialised

In 1862 a boy called John Merrick was born. As with any other baby, it had taken about 9 months for a fertilised egg to grow into the millions of cells that made up his body. This process involves not only cell division but cells becoming specialised for different jobs.

After John Merrick was born the process of growth continued in his body. However, before the age of two it was clear that something was wrong. Different parts of his body started to grow at an incorrect rate. This led to deformities over most of his body. He became known as the Elephant Man.

Extreme cases like that of John Merrick show how it is important it is for the process of growth to be carefully controlled by the body.

Growth problems rarely cause problems on this scale. However, after a human baby is born it is regularly checked to make sure that it grows and develops at the right rate. All babies develop at different rates but there are guidelines that show the range that is considered normal. The diagram on the right shows a chart used to plot the weight and head size of a baby boy.

Parents are encouraged to plot their baby's figures on the graph. It is quite usual for a baby's line, when it is plotted, to go across one of the lines on the graph. But if it drops suddenly this may be a sign of disease or growth problems.

John Merrick ▶

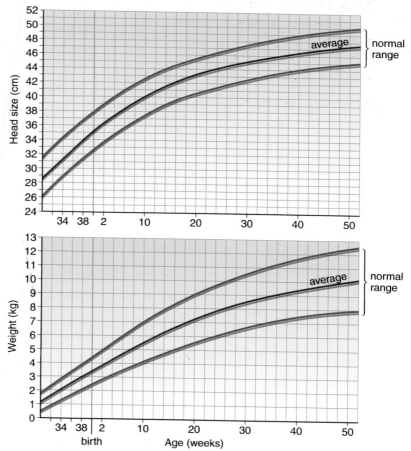

a Different graphs are used for boys and girls. Why do you think this is?

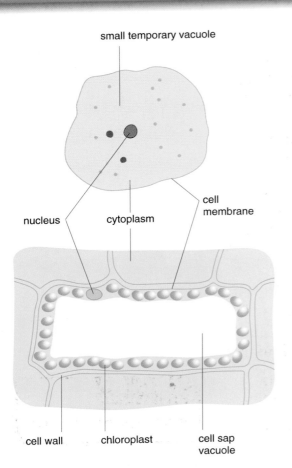

small temporary vacuole

nucleus cytoplasm cell
membrane

cell wall chloroplast cell sap
vacuole

▲ *Animal and plant cells*

Plant and animal cells

Plant and animal cells both contain certain important structures. These include the cell membrane, cytoplasm and nucleus.

There are, however, important differences. Only plant cells have:
- chloroplasts to absorb light energy for photosynthesis
- a cell wall to help support the plant
- a large vacuole that stores cell sap and also provides support.

Growth in plants and animals

Both plants and animals increase in size or grow during their lives. They do this by a combination of making new cells and by the new cells enlarging. However, there are some differences between growth in plants and animals.

Most animals tend to stop growing when they get to a certain size but plants can often carry on growing for the whole of their lives. Some plants can become very large indeed.

Plant cells also enlarge much more than animal cells after they have been produced. This is usually how plants gain height.

In plants, cell division tends to happen at the tips of roots and shoots. This means a plant develops a branching shape. In animals, cell division happens over the whole of the body.

b **How do animals and plants feed?**

c **How do you think their different shapes help them to feed?**

Differentiation and stem cells

Once new cells have been made by mitosis they take on different functions. This is called differentiation. In animals, the cells may become muscle cells, nerve cells, blood cells or any of the other types of cells that make up the different tissues and organs. In plants many cells take on different functions, but plants also keep a large number of cells that retain the ability to form different types of cells. You can show this because a small piece cut from a plant can grow into a whole new plant.

Modern research has shown that animals, including humans, do have some cells that still have the ability to differentiate. They are called **stem cells**. These stem cells are easy to find in the embryo but much harder to find in the adult body.

Growing babies

The length of time between fertilisation and birth is called the **gestation period**. During this time the zygote divides to form a ball of cells called an embryo. Once the embryo has formed all the organs and tissues of a baby, it is called a fetus.

Different animals have different gestation periods. The most important factor controlling this is the size of the animal. It takes about 22 months to grow a baby elephant but only about 3 weeks for a baby mouse!

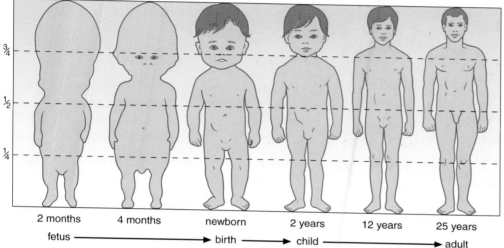

2 months   4 months   newborn   2 years   12 years   25 years

fetus ⟶ birth ⟶ child ⟶ adult

Different parts of the same baby grow at different rates. The diagram above shows that the head and brain of an early fetus grow very quickly compared with the rest of the body. Later, the body and legs start to grow faster and brain and head growth slows down.

Growing up

Humans grow quickly or slowly at different times in their lives. The graph on the right shows how human growth changes over a person's lifetime. During infancy a child grows very quickly. This slows down slightly during childhood but there is a growth spurt during adolescence (puberty). Once a person reaches adulthood they stop growing.

d   **Look at the graph. What are the differences between male and female growth patterns?**

e   **What happens to growth in old age?**

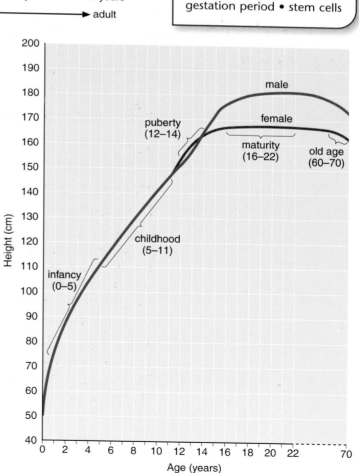

Uses of stem cells

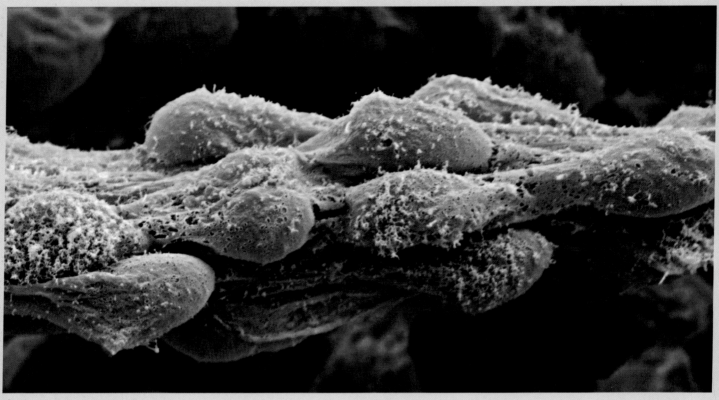

▲ *An embryonic stem cell*

There are two sources of stem cells. They can be obtained from an early embryo or from an adult body. Stem cells can be identified in an embryo when it is only about 5 days old. These embryonic stem cells can develop into any of the 200 tissues of the body.

Stem cells from an adult body are harder to find and often they can only form a limited number of types of cell. One good source of these adult stem cells is the blood found in the umbilical cord after a baby is born.

The most useful stem cells are embryonic stem cells. However, the embryos are destroyed at an early stage in order to remove the stem cells. This is very controversial. Scientists believe that these cells could be used to cure a large number of different diseases, including diabetes and spinal cord damage.

Newspapers often have articles about stem cell use:

Mother becomes pregnant so that baby's umbilical cord blood can be used to treat ill brother

Scientists plan to produce embryos to cure paralysed man

Questions

1 Explain why embryonic stem cells are more useful to scientists than adult stem cells at present.

2 Discuss why some people may be against the use of embryonic stem cells as described in the newspaper headlines.

3 Scientists think that they will be more successful if the stem cell donor is closely related to the patient. Suggest why they think this.

Growing around corners

In this item you will find out

- that plant growth is controlled by chemicals called hormones

- how these hormones control plant responses

- how these hormones can be used to change plant growth

If you have ever seen mushrooms growing out of a dead tree stump then you may have noticed that they always grow out sideways but then bend upwards. You can see this in the photograph on the right.

Growing this way ensures that the cap of the mushroom, where the spores are released, is always horizontal to the ground.

a **Why do you think this is so important for the fungus?**

Scientists decided to investigate how fungi managed to grow in this way. On one of the Spacelab missions they grew fungi in the Spacelab where there was zero gravity. The fungi grew completely differently from the way that they normally grow.

b **How does their growth differ on the Spacelab?**

c **Suggest a reason for this difference.**

Scientists now know that this type of growth is also shown in green plants and is controlled by chemicals. These chemicals are produced in the plant. They control many aspects of the plant's growth and development such as:
- growth of roots and shoots
- seed germination
- leaf fall
- disease resistance
- fruit formation and ripening
- flowering time
- bud formation.

Scientists have identified many of these chemicals and can now produce artificial versions of them. These are available to gardeners and farmers.

▲ *Fungi grown on Earth*

▲ *Fungi growing in the Spacelab*

95

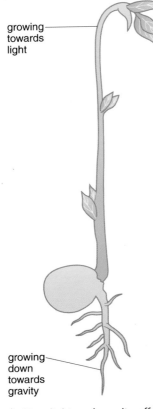

growing towards light

growing down towards gravity

▲ *How light and gravity affect plant growth*

Responding to light and gravity

Like other living organisms, plants must be able to respond to changes in their environment. This is called sensitivity. Animals often respond to stimuli by moving about but plants cannot do this. They respond by growing in a particular direction. Plant growth movements that are in response to stimuli from a particular direction are called tropisms and the growth is called a tropic response.

Two of the most important stimuli that plants respond to are light and gravity. A response to light is called a **phototropism** and response to gravity is **geotropism**. Different parts of the plant will respond differently to these stimuli. This is shown in the diagram on the left.

Shoots grow towards light so they are positively phototropic. Roots grow towards gravity and so grow downwards. They are positively geotropic.

d   **It is very important for the plant for the shoots to grow towards light. Why is this?**

e   **Why is it also important that the roots grow towards gravity?**

When a seed starts to grow in the soil the shoot grows upwards. This happens even though it is in the dark. So, as well as being able to grow towards light, shoots must also grow away from gravity.

Stimulus	Growth response	
	Shoots	**Roots**
Gravity	away = negatively geotropic	towards = positively geotropic
Light	towards = positively phototropic	away = negatively phototropic

Experiments have shown that these growth movements are controlled by chemicals in solution that can move through the plant. These chemicals are **plant hormones**. The main plant hormone in these responses is called **auxin**.

How do tropisms work?

Over the past century many scientists have studied plants to try to work out how tropisms work. The results of their experiments show that:

- auxin is made in the tip of plant shoots and diffuses down the shoot
- the auxin then causes the shoot to grow
- in uneven light more auxin is sent to the shaded side of the shoot causing the cells to grow more and elongate so that the shoot curves towards the light.

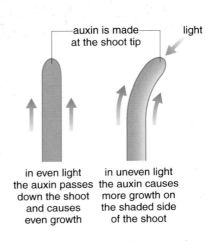

auxin is made at the shoot tip

light

in even light the auxin passes down the shoot and causes even growth

in uneven light the auxin causes more growth on the shaded side of the shoot

▲ *How plant auxin affects growth*

Uses of plant hormones

Scientists have developed many chemicals that are similar to auxin and other plant hormones. These can be used to aid plant cultivation and the effective production of food.

When they are sprayed on plants at high concentrations, they cause the plants to grow very quickly and die. Narrow-leaved plants such as grasses do not take up these chemicals and so the chemicals can be sprayed onto lawns. Because they will only kill the weeds, they are called **selective weedkillers**.

Other plant hormones encourage shoots to grow roots from their cut end. Gardeners use these hormones in **rooting powder** so that they can produce more plants by taking cuttings.

Plant hormones also control how quickly fruits ripen so they can be used to control fruit ripening – either slowing it down or speeding it up. This means that fruit such as bananas can be transported long distances and then be made to ripen at just the right time, ready to be sold in the shops.

◀ Bananas picked before they are ripe

Seeds, flowers and buds all start growing at the best time of the year to give the plant the best chance to survive. If conditions are tough they will not develop. This is called **dormancy**. Plant hormones can be used to control this dormancy so that particular plants and flowers can be made available all year round.

 f Why is it useful to be able to transport fruit such as bananas before they are ripe?

 g Why is it important for many flowers to open in the summer rather than in the winter?

> **Keywords**
>
> auxin • dormancy • geotropism • phototropism • plant hormone • rooting powder • selective weedkiller

To bend or not to bend?

The diagrams below show three famous sets of experiments carried out by different scientists. They all experimented on young plant shoots to try to find out more about tropisms. Use these diagrams to answer the questions.

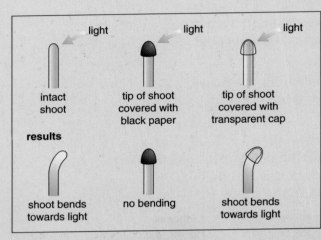

▲ Experiments by Darwin (1880)

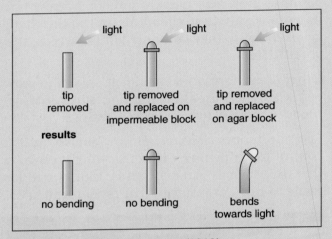

▲ Experiments by Boysen-Jensen (1913)

Darwin	English scientist Charles Darwin is famous for his work on evolution. He also carried out experiments on tropisms with his son Francis.
Boysen-Jensen	Peter Boysen-Jensen was a Danish plant biologist who became the first chairman of the International Plant Growth Association.
Went	Fritz Went was a Dutch biologist – he was the first to use the word auxin.

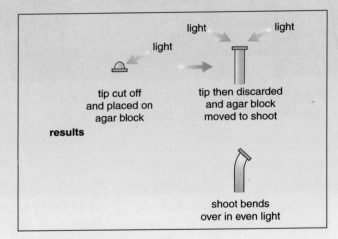

▲ Experiments by Went (1928)

Questions

1 Darwin's experiments showed where the light is detected in the plant. Explain how they show this.

2 Explain what conclusion Boysen-Jensen could make from his experiments.

3 The result of Went's experiment shows the shoot bending over although it is growing in even light. Explain why this is.

4 What would be a good control experiment for Went to have carried out?

Controlling the changes

In this item you will find out

- what causes mutations

- how selective breeding is used

- how genes can be moved from one organism to another

▲ *Albino rabbits are the result of a mutation*

Most people have heard of the word mutation and may describe organisms that are unusual as mutants. But what is a mutation and what causes it to happen?

In many science fiction films, strange organisms are often produced by mutation. This is often caused by leaks of radiation or unusual chemicals. The mutant seems to take on a new form and always seems to be dangerous. Is this really what mutation is all about?

There is some truth in these stories. Mutations are in fact changes in the genes of organisms which can happen spontaneously. As a result, these changes can alter how an organism looks or behaves.

Most mutations are harmful to the organism and many will kill it. However, occasionally a useful mutation occurs. This can allow types of organisms to change and become more advanced. We are all mutants – without mutations we would not be here!

Amazing fact

Every time a human gene is copied there is less than a one in a million chance that a mistake will be made.

 A mutation is more likely to have an effect if it occurs in a gamete. Suggest why.

Scientists could use chemicals or radiation to try to change the genes of organisms by causing mutations. However, the results would be random. Instead, scientists use two techniques to change organisms. They are **selective breeding** and **genetic engineering**.

Selective breeding has been used for thousands of years to produce organisms with certain characteristics. Most of our strains of farm animals have been produced in this way.

Now genetic engineering is being used because it produces quicker results. Both of these techniques have their critics and people are worried that scientists have taken these processes too far.

▲ *Maize can be genetically modified*

Mutations

Changes in the genes of an organism by mutation usually happen when the genes are copied before a cell divides. Under normal conditions, mutations usually happen at a very slow rate but the rate can be speeded up by:
- radiation such as X-rays or ultraviolet light
- certain chemicals such as the tar in cigarette smoke.

When a gene is copied, sometimes there is a change in the order of bases in the DNA. Since the order of bases codes for the proteins that a cell makes, this means that the cell may not be able to make the same proteins. This is why a mutation may change the cell and maybe the whole organism.

Selective breeding

If you look at all the different types of dogs that live today it is difficult to believe that they are all members of the same species. All the different types have been produced as a result of selective breeding. Two different breeds are shown in the photograph.

Some have been produced to be fast runners, others as guard-dogs and some as gentle pets. Selective breeding always works like this:
- two animals are chosen that have the characteristics that are wanted, such as speed
- these animals are allowed to mate – if they come from different breeds this is known as **cross-breeding**
- when the offspring are produced, the ones with the most desirable characteristics are chosen again and mated
- this happens for many generations until the animal with all the right characteristics is produced.

In this way humans have produced different breeds of dogs and champion racehorses.

Selective breeding can be carried out in the same way with plants. This means you can breed crop plants that are stronger, more resistant to disease or that are ready for harvesting earlier. This leads to improved agricultural yields.

Some people are worried that selective breeding uses organisms that are too closely related. This **inbreeding** can cause offspring to be born that have recessive abnormalities. It will also reduce the gene pool and there will be less variation in the population.

▲ Members of the same species

 Why might less variation in the population be a problem?

Genetic engineering

One of the problems with selective breeding is that it takes a long time to produce the organism that is needed. However, scientists can now choose what characteristics an organism has by changing its genes. Genetic engineering involves transferring genes from one organism to another. This will produce an organism that has different characteristics.

Bacteria can be genetically modified to produce human insulin which can be used by diabetics to control their blood sugar levels. It is also possible to genetically modify crop plants, such as maize or wheat, so that they become resistant to frost damage, disease or herbicides.

c   **What is the advantage of making crop plants resistant to herbicides?**

All of these examples of genetic engineering use a similar process. First, the scientists have to decide which characteristics they want to copy. Then the gene controlling these characteristics has to be found and isolated. The gene is cut out of the donor DNA using enzymes and is inserted into the host DNA again using enzymes. The host cell then copies itself (replication) including the new gene.

Keywords

cross-breeding • genetic engineering • inbreeding • selective breeding

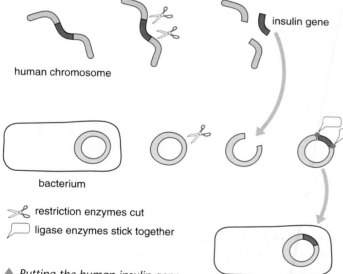

human chromosome

insulin gene

bacterium

✂ restriction enzymes cut

▭ ligase enzymes stick together

▲ *Putting the human insulin gene into bacteria*

Genetic engineering is useful because it produces new organisms quickly with the characteristics that we want. However, the new genes inserted into an organism may have harmful effects which the scientists were not expecting.

Some people are worried about this. They are concerned about eating food containing genetically modified crops and about the effects these crops will have on the environment.

d   **Why are some people worried about producing herbicide-resistant plants?**

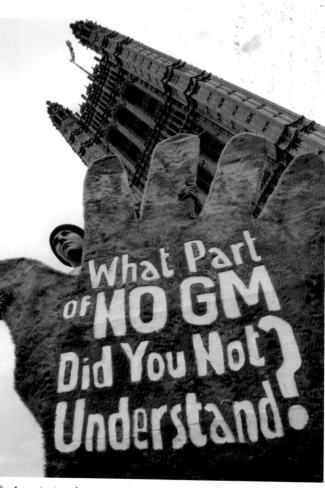

▲ *A protestor demonstrating against genetically modified crops*

Golden rice solves a problem?

In many parts of the world people eat a lot of rice and not many vegetable or dairy products. This means they can become deficient in vitamin A. This can lead to problems with their eyes. The World Health Organization estimates up to 500 000 children go blind each year because of vitamin A deficiency.

Vitamin A deficiency can cause this disease ▶

Vitamin A can be found in beta-carotene in carrots. Scientists have been able to take the genes from carrots that control beta-carotene production and insert them into rice. When people eat this genetically modified 'golden rice', they can convert the beta-carotene in the rice into vitamin A.

When the original strain of golden rice emerged from laboratories in Switzerland 5 years ago, it was hailed by some as an instant solution.

But not everyone believes golden rice is the best answer to Vitamin A deficiency. Some environmental groups say aiming for a balanced diet across the board would be a better solution. They point out that people who are deficient in vitamin A are also deficient in lots of other vitamins and minerals. They ask whether scientists are going to genetically modify a crop to solve each problem. They also say that not enough tests have been done on genetically modified crops.

Scientists have replied saying that for many years millions of pounds have been invested in improving people's diets but people are still suffering.

Questions

1 Describe briefly how golden rice can be made to produce beta-carotene.

2 Why do scientists think it is very important to produce this new rice?

3 Why are environmental groups against genetically modified rice?

Cloning around

In this item you will find out

- how plants and animals can be cloned

- the advantages and disadvantages of using cloned plants

- some of the benefits and concerns of using cloning

▲ *Dolly the sheep*

A **clone** is a genetically identical copy of an organism.

A breakthrough in cloning animals happened in the 1970s when John Gurdon managed to clone a frog using the cell of an adult frog. The trouble was, nobody could successfully do the same experiment on a mammal – that is until Dolly the sheep was born in 1996. So it was over 20 years after Gurdon's work that scientists managed to produce a clone of a mammal from an adult body cell.

Now cats, dogs and horses have all been cloned. But why should scientists want to do this?

People can pay companies to make a clone of their favourite pet so that its genes will be preserved forever. Farm animals that have the characteristics the farmer wants can be cloned to make large identical herds.

Animals could also be genetically engineered so that they contain some human genes. They could then be cloned and used to produce useful human proteins such as insulin. They could also be used as a supply of organs for human transplants.

a Why would a supply of animal organs for transplants be very useful?

b Why do you think it is better to transplant organs from animals that have human genes in them?

Amazing fact

Each year over 3000 people have an organ transplant in Britain.

It could be possible to ▶ have your pet cat cloned

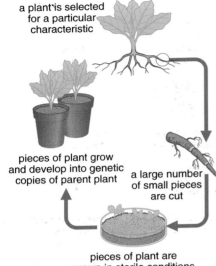

a plant is selected for a particular characteristic

pieces of plant grow and develop into genetic copies of parent plant

a large number of small pieces are cut

pieces of plant are grown in sterile conditions on growth medium

Cloning plants

Cloning is not new for gardeners. They have been using cloning technology for thousands of years. Some plants reproduce asexually and make clones of themselves. Many gardeners have used this process for centuries, but a more modern method uses **tissue culture**. This is useful because only a small quantity of tissue is needed rather than a large piece of a stem.

The scientists select a plant which has the characteristics that they want. They then cut the roots into lots of small pieces of tissue. The pieces of root are then grown in sterile conditions on a suitable growth medium. This is called **aseptic technique**. The pieces of plant grow into plants that are genetically identical to the original plant. This is shown in the diagram.

Producing plants by various cloning methods has some advantages and some disadvantages:

Advantages	Disadvantages
You know what you are going to get because all the plants will be genetically identical to each other and the parent.	The population of plants will be genetically very similar – there will be little variety.
You can mass-produce plants that do not flower very often or are difficult to grow from seeds.	Because the plants are all similar, a disease or a change in the environment could wipe them all out.

Cloning animals

Animals are much harder to clone than plants because, unlike plant cells, animal cells lose the ability to change into other types of cells (differentiate) at an early stage.

One way to clone an animal is to copy what happens in nature when identical twins are produced. An embryo is produced by a sperm fertilising an egg and then the embryo is split into two at an early stage. This is how cows can be cloned.

The scientists collect sperm from the bulls they have chosen. They then **artificially inseminate** selected cows with this sperm. When the embryos are large enough they are collected and split into two, forming clones. These embryo clones are implanted into cows which act as **surrogate** mothers and the calves are born normally. This process is shown in the diagram on the next page.

The problem with using this process to produce cattle is that you are never quite sure what you will get. Although the parent animals may be champion cattle, sexual reproduction is still involved. This means variation.

Dolly the sheep

Scientists tried for a long time to produce a clone of an adult mammal, without involving fertilisation. They succeeded in 1996 when a cloned sheep called Dolly was born.

Examiner's tip

Remember that most plants, such as strawberry plants, can also reproduce sexually by producing flowers.

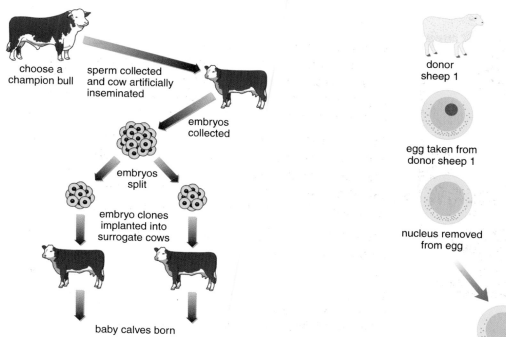

choose a champion bull

sperm collected and cow artificially inseminated

embryos collected

embryos split

embryo clones implanted into surrogate cows

baby calves born

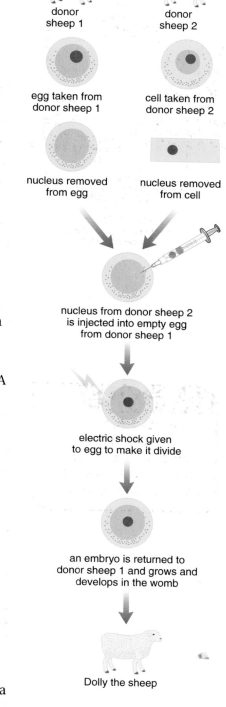

donor sheep 1

donor sheep 2

egg taken from donor sheep 1

cell taken from donor sheep 2

nucleus removed from egg

nucleus removed from cell

nucleus from donor sheep 2 is injected into empty egg from donor sheep 1

electric shock given to egg to make it divide

an embryo is returned to donor sheep 1 and grows and develops in the womb

Dolly the sheep

The scientists removed a nucleus from the egg cell of one sheep. They then replaced the egg cell nucleus with the nucleus from an udder cell from a second sheep.

As a result, the egg cell from the first sheep contained the nucleus and DNA of the second sheep. The egg cell was then implanted into the first sheep where it grew into a clone of the second sheep which had donated the udder cell. This is shown in the diagram on the right.

Should cloning animals be allowed?

Since Dolly was born in 1996, other animals such as pigs, cats, cows and horses have been cloned. Farmers might like the idea of having large herds of cloned prize-winning animals but some scientists believe that this could cause problems. If all the animals are genetically identical they could very easily be wiped out by a particular infection.

Using cloned animals as a supply of organs for human transplants would have important advantages but there are worries. Some people believe it is immoral and some are concerned that animal diseases could be spread to humans if this process was carried out.

However, there are other possible uses that are even more controversial. If animals such as sheep can be cloned then it should be possible to produce a cloned human embryo. This could be used in one of two possible ways:
- to extract stem cells at the embryo stage which could be used to treat illness (therapeutic cloning)
- to be implanted into a woman to produce a baby (reproductive cloning).

Keywords

artificially inseminate • aseptic technique • clone • surrogate • tissue culture

Human clones

The ability to clone humans does have some medical benefits. It could allow an infertile couple to have a child. Many couples cannot have children naturally. There may be something wrong with the man's sperm or the woman's eggs. An Italian doctor, Severino Antinori, wants to offer cloning as a treatment for infertility. A couple could have a baby that was a clone of the man or the woman.

Scientists could also obtain a supply of stem cells which would be taken from a cloned embryo. The embryo would then be destroyed and the stem cells could be used to repair damaged tissues. This technique has already been tried for treating rats that had been paralysed due to injuries to their spinal cord. Scientists hope that it can be used to treat humans that have spinal cord injuries.

Recently, some families have asked scientists if they would take some cells from a dead or dying child. The cells could be used to create a clone of a child that had been killed in an accident or had an incurable disease.

However, there are ethical dilemmas associated with human cloning. Some people believe that all human cloning is ethically unacceptable. They feel that killing an embryo, even at an early stage, is taking a life. They also feel that a cloned child would not be a true individual, as it would have exactly the same genes as another person.

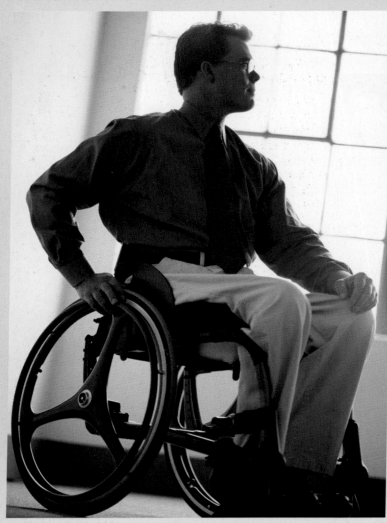

▲ *Stem cell research may help spinal injury victims*

Questions

1 Why might a couple prefer to use cloning to produce a baby rather than use an embryo donated by another couple?

2 If a clone of a dead child is produced, do you think the clone would be identical to the dead person in all their characteristics? Explain your answer.

3 Some people seem to think that cloning is acceptable if it is used to produce a supply of stem cells. However, many more people are against the idea of reproductive cloning. Why do you think this is?

B3a

1 *Where in a cell are mitochondria found and what is their function?* [2]

2 *Read the following statements about DNA.*

Which statements are true?
A *DNA controls the production of proteins.*
B *DNA consists of three chains wound together.*
C *The DNA molecule contains long chains of bases.*
D *DNA controls the production of amino acids.* [2]

3 *The diagram shows a molecule of DNA being copied (replicating).*

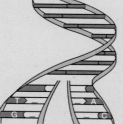

a *How can you tell that the molecule is replicating?* [1]
b *What type of chemical do the letters on the diagram represent?* [1]
c *What holds the two strands of the DNA molecule together?* [1]
d *Copy and complete the diagram by writing a letter in each of the four spaces on the diagram.* [2]
e *Explain why the order of these letters in the DNA molecule is so important.* [2]

4 *The graph shows the effect of pH on three enzymes.*

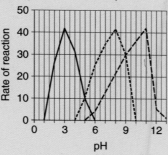

Key
—— Enzyme 1
------ Enzyme 2
----- Enzyme 3

Enzyme	Optimum pH
1	3.0
2	
3	

a *Copy the table (right) and use the graph to complete it.* [2]
b *Explain why enzyme 2 does not work above pH 10 or below pH 4.* [3]
c *Enzyme 1 is called pepsin. It breaks down protein. Explain why this enzyme can only break down protein and not starch or fats.* [2]

B3b

1 *The diagram shows part of an alveolus and a blood vessel.*

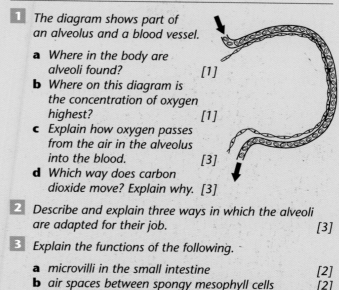

a *Where in the body are alveoli found?* [1]
b *Where on this diagram is the concentration of oxygen highest?* [1]
c *Explain how oxygen passes from the air in the alveolus into the blood.* [3]
d *Which way does carbon dioxide move? Explain why.* [3]

2 *Describe and explain three ways in which the alveoli are adapted for their job.* [3]

3 *Explain the functions of the following.*

a *microvilli in the small intestine* [2]
b *air spaces between spongy mesophyll cells* [2]
c *transmitter substances at the ends of neurones* [3]

B3c

1 *A table has been drawn to show some differences between arteries and veins.*

Copy and complete the table by filling in the blank boxes. [6]

Feature	Arteries	Veins
direction of blood flow		
type of wall		
are valves present?		

2 *Read the following information about some of the effects of smoking on the body.*

Two parts of tobacco smoke that damage the body are nicotine and carbon monoxide. Nicotine dissolves in the liquid part of the blood and increases the heart rate and blood pressure. This can be particularly dangerous if there are plaques present in the arteries. It also makes the platelets too active which can be dangerous.

Carbon monoxide combines with haemoglobin to form carboxyhaemoglobin which does not break down.

a *What is the name of the liquid part of the blood that nicotine dissolves in?* [1]
b *Explain how plaques form in arteries.* [2]
c *Suggest why might it be harmful if platelets become too active.* [2]
d *Explain how the combination of carbon monoxide with haemoglobin differs from the combination of oxygen and haemoglobin.* [2]

B3d

1 Explain why a sperm cell needs the following.

  a many mitochondria   [2]
  b an acrosome   [2]

2 The following diagram shows how sperms and eggs are produced. It also shows one sperm joining with one egg.

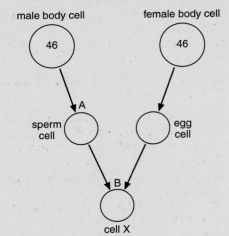

male body cell   female body cell

46   46

A

sperm cell   egg cell

B

cell X

  a The body cells have 46 chromosomes. What is the number of chromosomes in:
    i the sperm cell
    ii the egg cell
    iii cell X?   [3]
  b The type of cell division shown at A makes gametes. What is the name of this cell division?   [1]
  c Process B shows the gametes joining. What is the name of this process?   [1]
  d What is the name given to the type of cell labelled X?   [1]
  e The gametes are said to be haploid. What does this mean?   [1]
  f Why is it important that the gametes are haploid?   [2]

3 The diagram represents two animals. One is single celled and the other has eight cells.

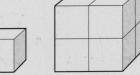

  a Each cell has dimensions of 1 × 1 × 1 unit. What is the surface area of each animal?   [2]
  b What is the volume of each animal?   [2]
  c Which cell has the smaller surface area : volume ratio?   [1]
  d Why is this a disadvantage to the cell?   [1]
  e If there are disadvantages for an organism if it becomes bigger then there must be some advantages. What are they?   [2]

B3e

1 Make a list of the structures that are found in **both** plant cells and animal cells.   [3]

2 Look at the diagram on page 21 showing the growth of different parts of the body.

  a Approximately what fraction of the body length is the head at birth?   [1]
  b Why do you think that the main growth of the head happens before the main growth of the arms and legs?   [2]

3 Read the following quote and answer the questions that follow.

'People are always against new ideas such as using stem cells. Within a few years they will be used all the time to cure diseases. It is not that people object to using stem cells, it is where they come from that worries some people.'
  a What are stem cells?   [2]
  b How can they be used to cure diseases?   [2]
  c Why do you think that some people are worried about where they come from?   [3]

B3f

1 The table shows some features concerning plant growth.

Copy and complete the table by putting a tick or a cross in each box to show whether they are true of shoots and roots.   [3]

Feature	Shoots	Roots
contain plant hormones		
show positive phototropism		
show negative geotropism		

2 Read the following passage and answer the questions that follow.

Plant hormones control how plants grow and so are often used in agriculture to alter plant growth. They might be used to change the rate that fruit ripens. This is particularly helpful with fruit such as bananas which have to be transported long distances. They can also be used as selective weedkillers because certain hormones will only kill the type of plants that have broad leaves.
  a Why is it helpful to change the rate at which fruit ripens?   [2]
  b What is a selective weedkiller?   [1]
  c Why can selective weedkillers be used on crops such as wheat and barley but not on tomatoes or lettuce?   [2]

3 The diagram shows two experiments on plant shoots.

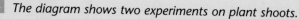

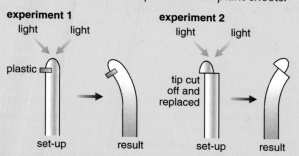

experiment 1

light   light

plastic

set-up   result

experiment 2

light   light

tip cut
off and
replaced

set-up   result

a What effect does auxin have on plant cells? [1]
b Explain why the shoot in experiment 1 bends to the left even though it is in even light. [3]
c Explain why the shoot in experiment 2 bends to the right. [3]

B3g

1 Copy and complete the following sentences.

A mutation is usually __(1)__ to an organism and can be caused by it being exposed to __(2)__ or __(3)__.

A more controlled way of changing the __(4)__ of an organism is by genetic engineering. This is now used to make organisms such as __(5)__ produce human insulin in large quantities. [5]

2 A type of cat has been bred that has very short front legs. These cats do not move around like normal cats but hop rather like kangaroos. This means that they cannot scratch with their paws or hunt animals.

a These cats have been produced by selective breeding. Explain the steps that would have been taken to do this. [3]
b Some people are keen to own these cats as pets. Others think that it is wrong to breed them. Describe the arguments that these two different groups of people might use. [4]

3 The diagram shows some of the stages in the production of human insulin by genetic engineering.

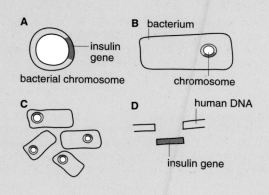

A

insulin gene

bacterial chromosome

B bacterium

chromosome

C

D

human DNA

insulin gene

a Arrange the four stages in the correct order. [3]
b Explain what is happening at each stage. [4]
c Suggest why some people are unhappy about the transfer of human genes into other organisms. [2]

B3h

1 Which pairs of organisms are clones?

A two strawberry plants grown from runners
B identical twins
C two strawberry plants grown from seeds
D a brother and sister [2]

2 Clones can be produced by splitting up embryos produced from fertilised cows' eggs.

a Write down the stages A to E in the correct order. [3]
   A cow artificially inseminated
   B sperm collected from champion bull
   C embryos collected from cow
   D cloned embryos put into surrogate cow
   E embryo split up
b Why do you think the farmers use a champion bull to provide the sperm? [2]
c What is a surrogate cow? [1]

3 Read the following passage about a gardener and answer the questions that follow.

George the gardener has a favourite geranium plant. He says that it is the same plant that was passed on to him by his father 30 years ago. Although it dies every winter he takes cuttings in the autumn and keeps them in a greenhouse for the next summer. This way he says he knows they will have attractive flowers.

a George says that his plant is the same plant that his father gave him thirty years ago. Why does he say this? [2]
b What does George do when he 'takes a cutting'? [3]
c How can George be so sure that the flowers will be attractive every year? [2]
d George has a whole flower bed full of these geraniums. Why does he have to be extra careful that they do not get a plant disease? [2]

4 Explain why the following statements are true.

a Cloning animals is much more difficult than cloning plants. [2]
b Tissue culture is a different process from taking cuttings. [2]
c It will not be easy to convince everybody that human cloning should be allowed. [2]

A walk in the country now is quite different from how it would have been 50 or 100 years ago. The fields are now much larger and farmers can use many different methods to increase their crop yields. These developments are possible because scientists now understand more about how plants produce their own food and grow.

Leaves are very efficient in trapping sunlight. As well as sunlight, plants need water and minerals. The water may have to pass many metres up to the leaves from the soil.

The minerals also come from the soil, and many are released from dead plant material. A study of how decay occurs can tell us a lot about how to stop the same thing happening to our food. All our methods of preserving food use this understanding.

I can't understand why anybody buys organic vegetables. They look just the same as the others but are much dearer.

I always buy organic vegetables and free range meat. I think that they are much safer to eat.

What you need to know

- Where diffusion occurs in plants.
- The word equation for photosynthesis.
- What a food chain shows.

Leaves for life

In this item you will find out

- what are the main parts of a leaf
- how a leaf is adapted for photosynthesis

Leaves come in many different shapes, sizes and colours, but the main job of leaves is photosynthesis. This process uses the green chemical, **chlorophyll**, to trap sunlight and produce food for the plant. Directly or indirectly all life on our planet depends on this function of leaves to supply food.

While most leaves are green, many can vary in colour and may have different colours on the same leaf. So how can some leaves be yellow, orange or red if they all contain chlorophyll?

Chlorophyll is a mixture of different coloured chemicals. This can be shown using the process of chromatography. A leaf is ground up to release the chlorophyll and then the different coloured chemicals are separated on special paper.

All plants have the green chemical, but some have red or yellow coloured chemicals as well.

In some plants leaves do extra jobs. The Venus flytrap plant has special leaves that trap unsuspecting flies if they touch the hairs.

 a How do you think the Venus flytrap plant digests the flies?

The leaves of the stinging nettle help to protect the plant. They have thousands of tiny hairs that are filled with poison. They break when touched and inject the poison into the skin.

 b What advantages do these hairs give to the plant?

Amazing fact

A nettle called 'devil-leaf', which is found in Papua New Guinea, produces a poison that is so strong it has killed people.

▲ *We can eat all these fruits and vegetables because of photosynthesis*

▲ *Venus flytrap*

▲ *Stinging nettle (magnified)*

The reactions of photosynthesis

The process of photosynthesis is carried out by all green plants and some bacteria. It traps the energy from sunlight and uses it to make food in the form of a type of sugar called glucose. The plant can then convert the glucose into all the other chemicals that it needs to live and grow.

The raw materials that are needed for the reaction are carbon dioxide and water. Fortunately for animals, the waste product that is given off in the reaction is oxygen which they can use for respiration.

The structure of a leaf

Although leaves come in many shapes and sizes, they all have certain things in common.

The diagram on the left shows the external features of a leaf. You can see that it has a series of **veins** that spread throughout it. They carry water that has been absorbed by the roots to the cells of the leaf. They also transport away the glucose that is made by photosynthesis.

When you look at a thin section of a leaf under a microscope, you can see details of the cells inside.

side vein

main vein

leaf stalk

▲ *The structure of a leaf*

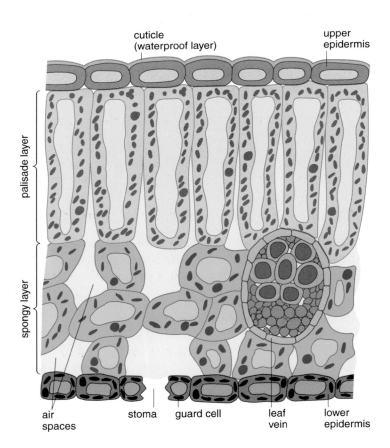

cuticle (waterproof layer)

upper epidermis

palisade layer

spongy layer

air spaces

stoma

guard cell

leaf vein

lower epidermis

Inside a leaf ▶

There is a waxy layer called the **cuticle** on the top of the leaf. Under the cuticle are different layers of cells. All the cells have a large central vacuole containing sap, as well as a tough cellulose cell wall. Some of the cells also have chloroplasts that contain the green chemical chlorophyll and so can carry out photosynthesis. The job of the chloroplasts is to absorb light energy.

 c What do you think the waxy cuticle is for?

 d Which types of cell contain chloroplasts?

On the bottom of the leaf are pores called **stomata** which are opened or closed by **guard cells**.

Photosynthesis efficiency

Although some photosynthesis occurs in plant stems, it is the leaves that do most of the work. Leaves have evolved over millions of years to make them efficient at carrying out photosynthesis:

- they are broad so that they have a large surface area to absorb light energy
- they are very thin so that gases do not have far to travel
- they have a network of veins that support the leaf, and carry water in and glucose out
- they have stomata to allow gases to pass in and out by diffusion
- they have chlorophyll to absorb sunlight
- the air spaces inside the leaf mean that the leaf has a large internal surface area-to-volume ratio.

▲ Leaves efficiently absorb light energy

The top layer of cells, the **upper epidermis**, has no chloroplasts and so is transparent. This means that light hitting the leaf passes straight through to the **palisade mesophyll** layer. This is close to the top of the leaf and contains most of the leaf's chloroplasts. It is where most photosynthesis takes place.

The air spaces in the **spongy mesophyll** layer allow gases to diffuse between the stomata and the photosynthesising cells. Carbon dioxide diffuses in and oxygen diffuses out of the stomata. All the thousands of cells inside the leaf give a very large surface area for the absorption of carbon dioxide.

Apart from the guard cells, the cells of the **lower epidermis** do not contain chloroplasts.

Keywords

chlorophyll • cuticle • guard cell • lower epidermis • palisade mesophyll • spongy mesophyll • stoma • upper epidermis • vein

Turning over a new leaf?

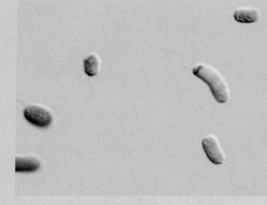

▲ *Purple sulfur bacteria*

All life on our planet depends on the leaves of green plants trapping the energy from the sun for photosynthesis. Scientists are now trying to use this process to produce fuel that could be used instead of coal or oil.

Scientists got the idea from the first organisms to be able to use photosynthesis – these were tiny bacteria. They lived millions of years ago in warm springs that are often found near volcanoes. Very similar bacteria can still be found in warm water in places such as Yellowstone National Park in America.

Instead of using water (H_2O) for photosynthesis like plants do today, these bacteria use the gas hydrogen sulfide (H_2S). The bacteria use this gas as a source of hydrogen so that they can make glucose from carbon dioxide.

Some time later, one type of bacteria stopped using hydrogen sulfide as their source of hydrogen. They started to use water instead. This produced oxygen as a waste product instead of sulfur. This small change was vital to life on our planet. It also meant that plants could live in many more habitats.

Scientists hope to be able to build 'artificial leaves'. These will be made of metal but containing some of the chemicals found in the cells of plant leaves or bacteria. These chemicals will allow the 'leaves' to use the energy from sunlight to split hydrogen sulfide or even water. This would produce hydrogen and maybe oxygen that could be used as fuel.

An advantage of these leaves is that they will not fall off in the winter!

Questions

1 Purple sulfur bacteria and green leaves both carry out photosynthesis. Explain the differences between the photosynthetic process in each.

2 Write down the word equation that purple sulfur bacteria use for photosynthesis.

3 Why could organisms that used water rather than hydrogen sulfide 'live in many more habitats'?

4 The artificial leaves, like real leaves, are likely to be flat sheets. Why do you think they will be this shape?

5 Artificial leaves might use some chemicals from the cells of the leaf that are used in photosynthesis. Which cells do you think these are?

6 Suggest why scientists are keen to develop new fuels such as hydrogen and oxygen.

Looking at osmosis

In this item you will find out

- what osmosis is

- about transpiration

- how plants try to reduce water loss

Gardeners know that if they want to move a plant then it is much better to take a large amount of soil with the roots. This helps to prevent the roots being damaged. When the plant is replanted it must always be watered to stop it **wilting**. This water is taken up into the plant from the soil by the process of **osmosis**.

To keep vegetables and salad crisp, a cook will often put them into water. This means that they will take up water by osmosis and be prevented from becoming limp.

But why do plants wilt if they do not get enough water?

Water moves in and out of plant cells through the cell wall and membrane. When the vacuole of a plant cell is filled with liquid, the water pressure presses up against the cell wall. The cell is said to be **turgid**. The cell wall is inelastic and does not stretch so it supports the plant cell. This turgor pressure is very important in supporting plants.

When the vacuole does not have enough liquid in it, because the plant is losing more water than it is taking in, it cannot press up against the cell wall and there is not enough turgor pressure. The plant cells become **flaccid**. This means they collapse inwards and the plant wilts.

Both the inelastic cell wall and water are needed to support plants.

Sometimes the cells may lose so much water that the cell membrane may come away from the cell wall. This is called **plasmolysis**.

a The cell wall is made of cellulose. What properties would you expect cellulose to have?

▲ This plant is wilting from lack of water

▲ This plant has plenty of water and its cells are turgid

Amazing fact

The forces created by seeds taking up water can be so large that ships' holds carrying seeds have been split apart when the seeds took up water and swelled up.

What is osmosis?

Osmosis is a special type of diffusion. Plant cells are surrounded by a cell wall and a cell membrane. The cell wall is **permeable** and so lets quite large molecules through. The cell membrane is **partially permeable**. This means that small molecules like water can diffuse through but larger molecules cannot.

Osmosis is the movement of water from an area of high water concentration (a dilute solution) to an area of low water concentration (a concentrated solution) across a partially permeable membrane. Like all diffusion it happens because the molecules are constantly moving about at random.

You can see how this works by looking at the diagram.

Because there is less water in the concentrated sugar solution, water passes into it from the weak solution. The level of the concentrated solution rises up the funnel as its volume increases. The dialysis or Visking tubing acts as a partially permeable membrane.

If plant cells are placed in water they take up water by osmosis. Under a microscope the cells can be seen to swell up and then stop. The liquid inside the vacuole is an area of low water concentration and the water moves into the plant cells. When the plant cells have taken up enough water, the cell wall stops any more water from entering.

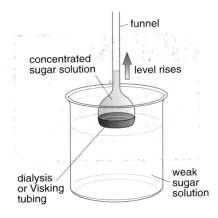

funnel

concentrated sugar solution

level rises

dialysis or Visking tubing

weak sugar solution

▲ How osmosis works

 b A plant cell with a vacuole containing cell sap equivalent to a 10% sugar solution is placed in a 5% sugar solution. Which way will water move by osmosis?

Animal cells and water

It is not just plant cells that take in water by osmosis – animal cells do too, but they behave differently from plant cells. When animal cells are surrounded by water they will take up too much water and burst. This is called **lysis**. The reason they burst is because animal cells do not have an inelastic cell wall to resist the pressure – they only have a cell membrane. In a concentrated solution, animal cells will shrink. This is called **crenation**.

> **Examiner's tip**
>
> When talking about osmosis it is best to talk about water concentration or sugar concentration. If you just say concentration the examiner might not know what you mean.

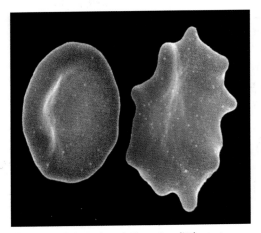

▲ *This cell has shrunk from too little water*

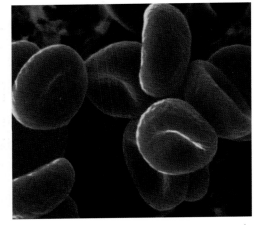

▲ *This animal cell has swollen from too much water*

A balancing act

We have already seen that plant leaves are adapted for efficient photosynthesis. They are covered in thousands of small pores called stomata that allow gases to diffuse in and out of the leaf. However, if gases can diffuse in and out, then water molecules will also be able to diffuse out. This loss of water is called **transpiration**.

Water is taken up from the soil into the roots by osmosis. The younger parts of the roots are covered in fine projections called root hairs. They increase the surface area of the roots to speed up the uptake of water.

 The roots grow from the tips pushing through the soil. Suggest why root hairs are only found on the younger parts of the root.

Although plants cannot stop losing water by transpiration, the flow of water to the leaves and into the atmosphere does help the plant in several ways:
- it helps to cool the plant down, rather like sweating in animals
- it provides the leaves with water for photosynthesis
- it brings minerals up from the soil
- it makes sure cells stay turgid for support.

Keywords

crenation • flaccid • lysis • osmosis • permeable • partially permeable • plasmolysis • transpiration • turgid • wilting

Cutting down water loss

Although transpiration has its uses, plants try to lose as little water as they can. They have a waxy cuticle on the top of the leaves which stops water leaving the leaf. They also only have a small number of stomata on the upper surface of the leaf. This cuts down on water loss because water is lost through evaporation from stomata when the sun's energy hits the leaf. Most of the energy hits the top of the leaf so if there are few stomata there then water evaporation is reduced.

Another way to reduce water loss is to close the stomata at night.

d Why is it less important for stomata to be open at night?

The stomata are opened and closed in a clever way by the guard cells on either side of the pores. When there is a lot of light and water available, they take in water and become turgid. This causes the cells to bend and the pores to open. When light levels are low, or when there is little water available, the opposite happens and the stomata close. This is shown in the diagram on the right.

Plants that live in areas where there is not much water available may have fewer stomata.

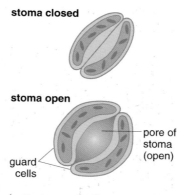

▲ How guard cells control stomata

Counting stomata

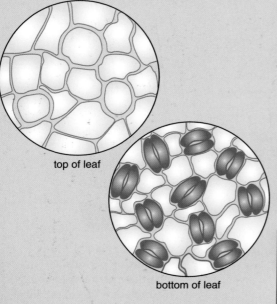

top of leaf

bottom of leaf

Many scientists think that counting stomata can tell us a lot about where plants live. It may also tell us about conditions in the past.

It is quite easy to count stomata in the lab. A leaf is painted with a small amount of nail varnish which is peeled off once it is dry. This carries with it an imprint of the stomata. They can then be counted under a microscope.

This technique can be used to work out the number of stomata on each surface of different leaves. The results of an investigation into holly and oak leaves are shown in the table.

Type of plant	Number of stomata (mm^2)	
	Top leaf surface	Bottom leaf surface
holly	0	113
oak	0	340

Scientists have tried growing the same type of plant in atmospheres with different concentrations of carbon dioxide. They have then counted the number of stomata on the leaves of the plants. They have found interesting differences in the numbers of stomata that the plants develop.

▲ Oak leaf

▲ Holly leaves

Questions

1. Why is it necessary to make an imprint of a leaf to view under a microscope?

2. The area of the field of view shown in the diagram is 0.05 mm^2. What is the frequency of stomata per mm^2 on this leaf surface?

3. Why do you think the holly and the oak have all of their stomata on the bottom surface?

4. Suggest why the holly has fewer stomata than the oak.

5. Describe and explain the pattern shown by the graph.

6. Suggest how scientists could use these findings to estimate carbon dioxide levels from hundreds of years ago.

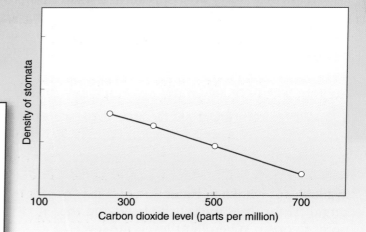

The scientists think that they might be able to use this discovery to find out about atmospheric carbon dioxide levels from many years ago.

Xylem and phloem in action

In this item you will find out

- about xylem and phloem
- what alters the rate of transpiration

Small insects called aphids feed on the stems of plants by inserting their long, needle-like mouthparts straight into special cells in the stem. It is these cells that transport sugar around the plant. The aphids don't even need to suck – the sugar just oozes out.

▲ Aphid sucking sugar

Plants transport many different substances both up and down their stems. As well as the products of photosynthesis from the leaves, there is also water which comes from the soil. But how do plants move these substances around?

If you cut a celery stem and put the end in a beaker of water which has been dyed red, then leave the celery for an hour, the water and dye move up the stem. If you cut a section through the stem, you will see that the water does not move throughout the stem but only moves in certain cells.

By giving the aphids some anaesthetic and then breaking them off, leaving their mouthparts in the cells, scientists have located the cells that move sugar. These cells are different from the water-transporting cells.

This can be confirmed if a plant is provided with carbon dioxide containing radioactive carbon. The plant will make radioactive sugar. When scientists take a section through the plant's stem the radioactive sugar can be detected. The sugar can be found close to the cells where the water is transported, but the sugar is in different cells. So plants transport different substances around in different ways.

Amazing fact

A female aphid can give birth to 12 babies a day, every day!

▼ Red dye travels up a celery stalk

a The water containing the dye and the radioactive sugars are in different cells. Which direction is each moving in the plant?

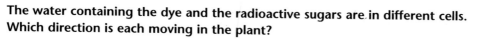

Xylem and phloem

The two different transport tissues in a plant are called **xylem** and **phloem**. In the diagram, you can see how they are arranged in the roots, stem and leaves of a **dicotyledonous** plant (it has two seed leaves).

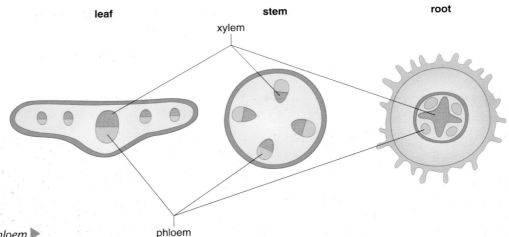

leaf   stem   root

xylem

Xylem and phloem ▶   phloem

Xylem transports water and dissolved minerals. The water and dissolved minerals are taken in by the roots and move up to the leaves and shoots where the water is lost in transpiration.

Phloem transports dissolved food substances (sugars) around the plant. Unlike the transport of water, the food may be transported up or down the stem. It may move from the leaves where it is made up to the tips of the shoots where growth happens. It may also be sent down the stem to storage organs such as swollen roots. This movement of food substances by the phloem is called **translocation**.

b   **What is the sugar used for in the tips of the roots and shoots?**

The xylem vessels and phloem tubes run continuously from the roots to the stem and into the veins of the leaves. In the stem they are gathered together into collections called **vascular bundles**.

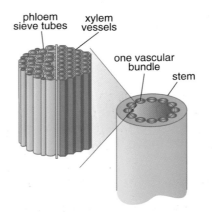

phloem sieve tubes   xylem vessels

one vascular bundle

stem

Vascular bundles ▶

Specialised cells

The cells that make up xylem and phloem are specialised so that they can carry out their functions.

Xylem contains long thin tubes called vessels each with a hollow **lumen**. This means they are made from dead cells. The vessels are hollow so that water can easily pass through. Their cellulose cell wall is thickened to stop them collapsing and so xylem also helps to support the plant.

Phloem is made up of columns of living cells. The cells have holes in the end walls to allow the food to pass through.

 Moving food through phloem requires a large amount of energy. Explain why it is important that the cells making up phloem are alive.

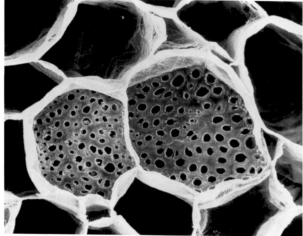

▲ Phloem tubes

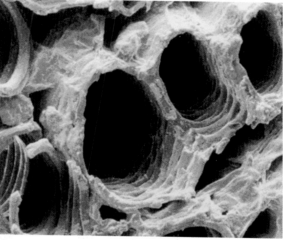

▲ Xylem vessels

Transpiration rate

As we have already seen, transpiration is the diffusion of water out of leaves through the stomata and its evaporation. When water is lost through the stomata, this creates a suction force that helps to pull more water up through the xylem vessels from the roots to the leaves.

The rate of transpiration is increased when:
• the temperature increases
• light levels increase
• it is windy
• it is dry and not very humid.

This is because, in warmer conditions, the water molecules have more energy and so evaporate faster. Also, an increase in light will cause more of the stomata to open and more water to diffuse out, while in windy conditions the water vapour in the air is blown away allowing more of it to evaporate. Finally, in dry conditions, the air holds fewer water molecules and so more water diffuses out of the leaf.

 Suggest why gardeners do not dig up and transplant plants on a hot, sunny day.

Keywords

dicotyledonous • lumen • phloem • translocation • vascular bundle • xylem

Measuring water uptake

Dr Steven Okinwe is measuring water uptake by a leafy shoot using a device called a potometer.

He cuts off a leafy shoot from a plant underwater by holding the stem in a bucket of water. He then inserts it into the potometer underwater in a sink.

All the joints of the apparatus are greased to stop leaks and all air bubbles are removed.

As the shoot loses water, it draws water up from the capillary tube. The movement of the water along the capillary tube can be measured every minute. This can be done under different conditions.

His results are shown in the table.

plant takes up water

reservoir

air is taken in →

capillary tube   water

	Distance water moves from start position (mm)							
Conditions	0 min	1 min	2 min	3 min	4 min	5 min	6 min	7 min
leaves uncovered	0	3	7	10	14	17	20	23
top of leaves covered in grease	0	2	5	8	11	15	18	21
bottom of leaves covered in grease	0	1	2	4	6	7	9	10

Questions

1 Write down two of the precautions taken when setting up the apparatus and explain why they are necessary.

2 The apparatus does not measure water loss from the shoot but measures water uptake by the shoot. Why does the shoot take up slightly more water than it loses?

3 Plot the results on a grid. Use the same grid for all three conditions.

4 Explain the results of the experiment as fully as possible.

A healthy diet

In this item you will find out

- what plants use minerals for
- what happens to plants if they lack minerals
- how plants take up minerals

Just like us, plants are in desperate need of **minerals** so that they can grow properly.

The difference between humans and plants is that we get our minerals from food but plants get their minerals from the soil. All plants can make their own food by photosynthesis but they need minerals in order to grow properly and produce flowers and seeds.

The minerals that plants need are absorbed from the soil by the roots. They are dissolved in water in quite low concentrations. Minerals are so important to plants that they go to great lengths to get them. They spread their root systems out wide in order to get as many minerals as possible from the soil.

Some plants recruit assistance in order to absorb minerals from the soil. Fungi are very good at taking up minerals even when they are in small concentrations. Some plants have fungi permanently attached to their roots. They spread out and take up minerals, passing some on to the plant. Lichens are combinations of fungi and single-celled algae. The fungi absorb water and minerals and share these with the algae.

Other plants that grow in soil which is very low in minerals catch insects and extract the minerals they need from them. They are often called carnivorous plants. They include the Venus flytrap, pitcher plants and sundews.

a How do you think these plants catch insects?

▲ *Lichens consist of fungi and algae*

Amazing fact

Plants are usually said to respond slowly to stimuli but the Venus flytrap can shut its trap in less than 0.1 of a second.

Pitcher plant ▶

Minerals

Plants can make sugars in photosynthesis using carbon dioxide and water. However, to turn these sugars into other important substances, such as protein and DNA, they need minerals. This is because sugars only contain the elements carbon, hydrogen and oxygen. Other chemicals that plants need contain elements such as nitrogen, phosphorus, potassium and magnesium. The table shows how plants obtain and use these elements.

Element required	Main source	Used by plants to produce
magnesium	magnesium compounds	chlorophyll for photosynthesis
nitrogen	nitrates	amino acids for making proteins which are needed for cell growth
phosphorus	phosphates	DNA and cell membranes to make new cells for respiration and growth
potassium	potassium compounds	compounds needed to help enzymes in photosynthesis and respiration

Examiner's tip

Do not call minerals 'food'. Plants make their own food by photosynthesis. They do not get food from the soil.

Mineral deficiencies

Just like people, if a plant grows without enough minerals it will not develop properly. This is called a **deficiency**. It is possible to demonstrate the effect of a lack of mineral by growing plants in solutions that are missing particular minerals. This is shown in the photograph below.

COMPLETE | −N | −P | −K | −Ca | −Mg | −S | −Fe

Bean plants with mineral deficiencies ▶

Different mineral deficiencies can affect plants in different ways. Some of these are shown in the table.

Mineral	Effect of deficiency on growth
magnesium	yellow leaves
nitrate	poor growth and yellow leaves
potassium	poor flower and fruit growth and discoloured leaves
phosphate	poor root growth and discoloured leaves

b Look at the two tables. Why do you think a lack of magnesium leads to yellow leaves?

c Suggest why a lack of nitrates lead to poor growth.

Fertilisers

Farmers may add fertilisers to the soil to help their crops grow. These fertilisers contain various combinations of minerals, such as nitrates, potassium, magnesium and phosphate compounds. This is because some plants need more of one type of mineral than another.

The photograph shows a packet of fertiliser with an NPK value on it. This stands for nitrate, phosphate and potassium and tells the farmer the proportion of each of these minerals in the fertiliser.

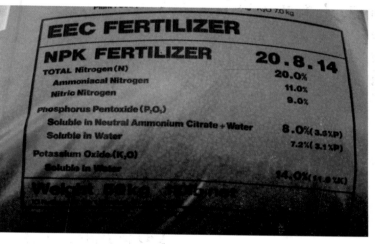
▼ Different fertilisers contain different proportions of minerals

Absorbing minerals

The minerals that the plant needs are absorbed from the soil by the roots. They are dissolved in water in quite low concentrations.

The concentration of these minerals in the root hairs is much higher. This means that they cannot be taken in by diffusion. A process called **active transport** is needed. This process moves substances from low concentration to high concentration. This means that it moves substances against the **concentration gradient**. It requires the energy from respiration to do this.

Because active transport uses energy from respiration, roots need oxygen in order to take up minerals by this process. Farmers try to make sure that their soil is not waterlogged because this reduces the oxygen content of the soil.

d Explain why there is likely to be less oxygen available to the plant roots in a waterlogged soil.

Amazing fact

The water that a giant sequoia tree takes up in 24 hours contains enough minerals to pave five metres of a four-lane motorway.

Keywords

active transport •
concentration gradient •
deficiency • mineral

▲ *Mycorrhiza help plants*

The fungal helpers

Not many people have heard of mycorrhiza but about 95% of all plants have them. They are fungi that live attached to the roots of plants. They spread out in the soil and allow the plant to absorb minerals at a quicker rate. They can be seen in the photograph on the left.

It is thought that the mycorrhiza are particularly good at absorbing phosphates from the soil that plants find difficult to take up.

All this is theory but scientists have carried out experiments to test these ideas. Small plants were grown in normal soil and in soil that had been sterilised. The sterilising kills the fungi. After some time, the mass of the plants was measured. The mineral content of each group of plants was also measured.

The results are shown in the table.

Conditions	Dry mass of plant (g)	Mass of mineral in plant (mg)		
		Nitrogen	Phosphorus	Potassium
no fungi present	0.3	2.5	0.2	2.0
fungi present	2.8	37.5	4.0	36.0

Scientists have used this type of evidence to show that plants gain significantly from their relationship with the fungi. But as the famous saying goes 'there is no such thing as a free lunch' – the fungi must be getting something in return from the plant.

Questions

1 The mycorrhiza is a fine network of threads of fungus attached to the roots. Suggest how this helps the roots to absorb more minerals.

2 How much more nitrogen is there in the plants with fungi compared to the plants without?

3 Suggest how this helps to explain the improved growth rate of these plants.

4 It is thought that fungi are particularly good at absorbing phosphates. Does the experiment show any evidence for this?

5 The fungi may be getting something in return from the plant. Suggest what this could be.

Pass it on

In this item you will find out

- about pyramids of numbers and biomass

- how energy flows through a food chain

- how we can use biomass

Plants can trap the energy from sunlight and use it to produce many products that are useful to us. Some products are used for food by animals although sometimes it can be difficult to make use of the energy trapped. For example, celery has energy trapped in its cells but it takes more energy to digest the celery than is released.

Other plants, such as the macadamia plant, produce material that is very high in energy. The nut of the macadamia plant contains almost 100 times more energy than celery but the nut is enclosed in the hardest shell in the world! Special nutcrackers are needed to break them open.

As well as using plant products for food, man also uses them for fuel.

Humans have burned wood for thousands of years but now there are many other ways that man is trying to use plants for fuel.

▲ Eating celery takes energy!

One idea has been put forward by a group of Japanese scientists. They want to build 100 vast nets that will float in the middle of the ocean. On the net will grow a quick-growing seaweed. The nets will then be towed back to land and the seaweed harvested. It will be dried and burnt to produce electricity.

▼ A burning log fire

a Suggest one advantage of burning the seaweed for energy rather than burning coal or oil.

Amazing fact

An aircraft carrier travels about 5 cm on each litre of fuel.

Pyramids of numbers and biomass

You can see how organisms rely on each other for food if you draw a simple food chain like the one in the diagram below.

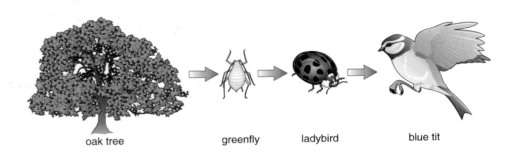

oak tree   greenfly   ladybird   blue tit

In every food chain the first organism is the producer. This means that it can make its own food using the energy from sunlight. All the other organisms in a food chain are consumers. They need to take in food because they can't produce their own.

To give us more information about the numbers of organisms in a food chain you can construct a **pyramid of numbers**. The number of organisms at each stage in the food chain (**trophic level**) is counted. Each box in the pyramid is drawn so that the area represents the number of organisms.

The trouble with this type of pyramid is that it does not take into account the size of the organism. One oak tree takes up as much area as one greenfly!

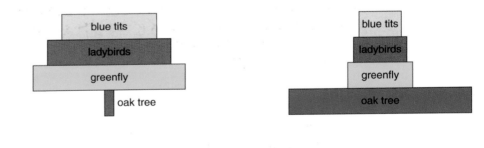

▲ Pyramid of numbers   ▲ Pyramid of biomass

An alternative is to draw a **pyramid of biomass**. Biomass is the mass of living material of an organism. The mass of all the organisms at each level is measured and the boxes are drawn to show the mass at each stage in a food chain or web.

b Why is the oak tree box much larger in the pyramid of biomass?

c Why do you think it is harder to get the information needed to construct a pyramid of biomass?

Energy flow

Some scientists study the flow of energy through food chains. The energy enters the food chain when plants absorb sunlight. The producers trap some of this energy by photosynthesis and convert it into chemical energy in compounds such as glucose. This energy then passes along the food chain as each organism feeds on other organisms and takes in the compounds. This energy flow is shown in the diagram.

The diagram on the right shows that energy is leaving the food chain at each stage. This is because organisms give out heat that has been made in respiration. Some energy is also lost in material that is ejected from animals (**egestion**). This is food that has passed all the way through an animal and has not been digested.

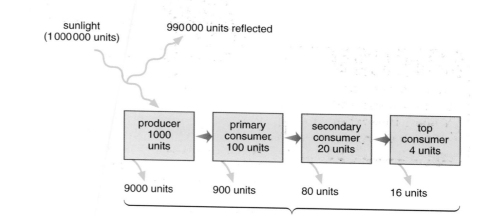

▲ *Energy flow in a food chain*

The efficiency of energy transfer explains the shape of a pyramid of biomass. Due to energy loss, a larger mass of organisms is needed to support the level above. This results in a pyramid. The loss of energy also explains why food chains rarely have more than about five levels. Because of the energy loss at each stage, not enough energy is left to support any more levels.

d Look at the diagram showing energy flow through the food chain. (i) What percentage of the energy is transferred from the producer to the primary consumer? (ii) What percentage is then transferred from the primary consumer to the secondary consumer?

Energy from biomass

We can use the energy stored in biomass in different ways:
- we can burn wood that comes from fast-growing trees
- we can produce alcohol by using yeast to ferment the biomass
- we can produce biogas which contains gases produced by bacteria fermenting the biomass.

All of these types of biomass are called renewable. This is because they can be produced at the same rate as they are used so they will not run out. **Biofuels** help countries that do not have reserves of fossil fuels so they do not need to import gas or oil. An added bonus is that when they are burned, the biofuels only release the same amount of carbon dioxide as is taken in to produce them. The carbon dioxide levels in the air should stay constant so there is no increase in air pollution.

Of course, the energy stored in biomass can also be used by humans or livestock by eating it, or we can grow the seeds into new plants to use as fuel. .

> **Keywords**
>
> biofuel • egestion • pyramid of biomass • pyramid of numbers • trophic level

Nuts to energy

▲ *Macadamia nuts*

In Australia, the macadamia nut is in great demand for baking and sweet making. There are now more than 4.5 million macadamia trees growing in the country. The problem is waste. The nut has a very hard and thick shell. One factory can produce about 10 000 tonnes of waste shells each year.

There is also pressure from the Australian government to generate more energy from renewable resources. Strict penalties apply if businesses do not do this.

Now one company has come up with an answer that will solve both of these problems – burn the nutshells!

A nut company has teamed up with an Australian energy company and built a large power station. This will be powered by burning the nutshells. It should generate up to 9.5 gigawatt hours which is enough to power about 1200 homes. Some of the power will be used to supply the nut factory but the rest will be sold to the national grid.

It has been estimated that burning the nuts rather than fossil fuels will reduce carbon dioxide emissions by about 9500 tonnes each year. This is the same as taking 2000 cars off the road.

A spokesperson for the company said that if successful this idea could be used in a range of other industries, including peanut, timber, wheat and grain processors.

Questions

1 What is the problem with the macadamia nut industry?

2 The Australian government wants energy to be generated from renewable fuels. Suggest why the nutshells are considered to be a renewable energy source but coal is not.

3 Burning the shells releases carbon dioxide. However, burning the nuts rather than fossil fuels will cut down on the release of 9500 tonnes of carbon dioxide a year. How do you think this reduction of carbon dioxide is possible?

4 Suggest why the Australian government is so keen to reduce carbon dioxide output.

Food for everybody

In this item you will find out

- some of the advantages and disadvantages of intensive farming
- how plants can be grown without soil
- about organic farming

▲ Traditional ploughing with oxen

Up until the last century farming methods had changed little for hundreds of years. Traditionally, farmers ploughed small fields using animals. Seeds were planted and crops harvested by hand.

In the last 500 years the world's population has doubled nearly four times. This increase in numbers has meant that there is a much greater demand for food.

This has resulted in major changes to farming methods. Crops are grown in large glasshouses and fish are bred in fish farms so that they grow much more quickly. Machinery has been developed to help the farmer. More and more chemicals are also available to farmers to use as fertilisers and to kill pests. This is known as **intensive farming**. This means trying to produce as much food as possible from a certain area of land and from the plants or animals that are farmed. In some areas crop yields have gone up tremendously because of intensive farming.

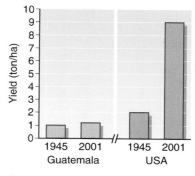

▲ Crop yields in Guatemala and the USA

The graph on the right shows the yield of one crop, corn, in the USA and Guatemala. Intensive farming techniques are used in the USA but not in Guatemala.

a How does the yield compare between the two countries?

b By using intensive practices, the USA has increased the yield of corn that is grown. By how much has the yield increased in 2001 compared with 1945?

Battery farming involves keeping animals, such as chickens, in controlled conditions indoors. They are kept warm and their movements are restricted. This means that they will lose less energy as heat.

Some people think that raising animals in this way is unethical. Chickens, for example, have very little room to move around in and because they grow so quickly, their legs often cannot support their bodies.

Amazing fact

Sales of organic meat in the UK went up by 139% between 2001 and 2004.

▲ Fish can be intensively farmed

BATTERY FARMING

Energy efficient

Producing food by intensive farming improves the efficiency of energy transfer along food chains in several ways. Intensive farmers often use **pesticides**. These are chemicals that kill pests. Removing pests that eat part of the crop leaves more biomass and energy for farmers to harvest.

Farmers may also use herbicides to kill plants, such as weeds, in crop fields. This helps to prevent competition from the weeds and so the crops can more efficiently trap the energy from the sun.

c Suggest three things that the crop plants and weeds are competing for.

d Why do you think farmers spray herbicide rather than weeding by hand?

Damaging effects

Although using pesticides and herbicides has increased food production, there have been drawbacks.

Pesticides may harm useful organisms such as insect pollinators. They may also enter the bodies of organisms. As they do not break down, their concentration may build up in consumers higher up the food chain.

e Look at the diagram on the left. Suggest why the fish do not die from the pesticide.

Soil free?

Many farmers that use intensive farming grow plants in glasshouses. This helps to protect the plants from extremes of weather. Sometimes they are grown in soil but not always. The growing of plants without soil is becoming more popular in many areas.

This is known as **hydroponics**. Proper hydroponics means growing crops in water, but farmers may use an artificial soil. This technique may be very useful in areas where the soil is poor (or barren, as it is known), or for growing plants like tomatoes in glasshouses.

The disadvantages of hydroponics are that the plants are not supported by deep soil and you need to add fertilisers to the water or artificial soil. The advantages are that the mineral levels can be carefully controlled and you can control diseases by adding pesticides to the water so that they reach each plant.

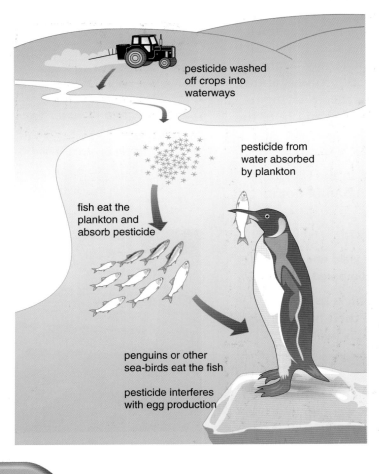

pesticide washed off crops into waterways

pesticide from water absorbed by plankton

fish eat the plankton and absorb pesticide

penguins or other sea-birds eat the fish

pesticide interferes with egg production

◄ *How pesticides travel up a food chain*

No artificial additives

Due to the problems caused by intensive farming, lots of people think that plants should be grown without artificial fertilisers, herbicides or pesticides. This is called **organic farming**.

There are a number of alternative methods that organic farmers can use:

- animal manure and compost can be dug into the soil as fertiliser
- crops that can fix nitrogen in the soil, such as clover, can be grown
- crop rotation can be used so that the same plant is not grown in the same field each year and pests cannot build up in the soil.
- crops can be weeded by hand
- farmers can vary seed planting times.

All these methods may be more labour intensive but they reduce the need to use chemicals that some people believe may be harmful.

The success of organic farming may depend on people being prepared to pay slightly higher prices for their food. In the developed world this may be possible but in the developing world it may be harder to convince farmers that they should not use intensive methods.

Pest control

Instead of using pesticides and insecticides to control pests it is possible to use living organisms. This is called **biological control**. Often the organism used is a predator that eats the pest. For example, farmers growing crops in glasshouses can buy packets of spiders. These spiders are released in the glasshouse and they eat a pest called the red spider mite.

Care must be taken when biological control organisms are introduced because they may have effects on the food web. If they wipe out the pest completely then this may mean that other animals in the food web may starve and die out.

Sometimes the control organism or other animals may increase in numbers and become pests themselves.

▲ Tomatoes growing using hydroponics

Keywords

battery farming • biological control • hydroponics • intensive farming • organic farming • pesticide

◀ A ladybird eats an aphid

▲ Prickly pear

▲ Cane toad

You win some and you lose some

Many of the advantages and disadvantages of biological control have been learnt in Australia. This is because Australia has been an island for a long time and has an unusual selection of plants and animals living there. This means that any new organisms that are introduced can rapidly become pests. The first real success was controlling a plant pest called the prickly pear.

This plant grows naturally in countries such as Argentina but was introduced into Australia by man and soon spread. By 1925 it was completely out of control, spreading at the rate of half a million hectares a year.

In 1925 massive amounts of chemical poisons were used to try to kill the prickly pear but it was still spreading. The Australians had to do something so they introduced a caterpillar from Argentina called cactoblastis. After careful testing the caterpillars were first released in 1926. They started to eat their way through the prickly pears and within six years the plant was under control.

But there have been failures. Fresh from the prickly pear success, the Australians introduced 102 cane toads in 1935. The idea was for the toads to eat cane beetles, which were pests of sugar cane. The trouble was, the toads eat other food as well. They reproduce really quickly and produce poison from glands on their backs. There are now millions of the toads in Australia and they are still increasing in numbers, threatening native frogs.

Questions

1 What is meant by biological control?

2 Why do you think that the Australians looked in Argentina for a control animal for the prickly pear?

3 Why was cactoblastis tested before it was released?

4 Cactoblastis only eats prickly pear but the cane toad eats many different foods. What lessons do you think scientists have learnt from this?

5 Write down one advantage and one possible disadvantage of biological control using these examples from Australia.

To rot or not to rot?

In this item you will find out

- what affects the rate of decay
- the type of organisms that cause decay
- some of the methods used to preserve food

Plants and animals are all made from organic material. When this material dies it starts to break down or **decay**.

In September 1991, two people walking in the mountains near the border between Austria and Italy made an amazing discovery. Half buried in the ice was the body of a dead man. They thought that he had died recently of an accident or even murder. When people looked more closely at the body they found that he had a copper axe. They soon realised that this body was very old – in fact it turned out to be 5300 years old!

▲ Otzi the iceman

The preservation of the iceman is an extreme example. However, in the case of our food, most of it has to be preserved so that it can reach our tables without breaking down first. Prawns are a good example – they last 10 days if kept at 0°C, but only 2 days at 10°C.

All this shows that dead animals and plants break down very easily. Sometimes we want to speed up this process. A gardener wants dead plants to break down quickly to make compost so provides the best conditions for this by building a compost heap.

But in the food industry we want to slow down or prevent decay. This may involve controlling the temperature that food is kept at but there are many other methods. The aim is to make sure that the food does not deteriorate before we eat it.

▲ Seafood kept on ice at a fishmonger's

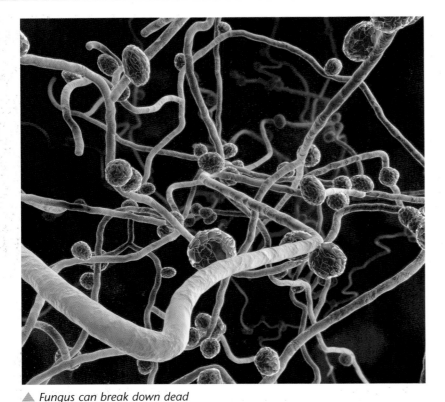

▲ *Fungus can break down dead organic material*

Organisms that cause decay

Organisms that break down dead organic material are called decomposers. They are very important because they allow chemical elements to be recycled. If decomposers did not do this all the chemical elements needed for life would build up inside dead organisms.

The two main groups of decomposers are bacteria and fungi. They release enzymes on the dead organic material and then take up the partially digested chemicals.

This type of feeding is called **saprophytic nutrition** and the bacteria and fungi are called **saprophytes**.

There are organisms that help the decomposers to do their job. Animals such as earthworms, maggots and woodlice feed on pieces of dead and decaying material (**detritus**). They are called **detritivores**.

Detritivores increase the rate of decay by finely breaking up material so it has a larger surface area. This means that it can be broken down faster by the decomposers.

 a Suggest why a piece of apple that has been dipped into disinfectant would decay faster than a piece that had not been treated.

b Suggest what effect temperature would have on the rate of decomposition caused by the saprophyte's enzymes.

Rate of decay

In order for organic material to decay, several things need to be present: microorganisms, oxygen and water. Oxygen is needed for the aerobic respiration of the microbes, while water is needed to allow substances to dissolve and the chemical reactions of respiration to occur. It also needs to be warm enough.

◀ *Worms are detritivores*

The rate of decay can be changed if the temperature changes, or if there is a lack of oxygen or water. If it is too hot or too cold, too dry or lacking in oxygen, decomposition will not occur. This is because the respiration and growth of the microbes will be slowed down.

Preserving food

The food that we eat is organic material and so is a target for decomposers to break down. To prevent this happening we use different techniques to reduce the rate of decay. This is called **food preservation**. Most food preservation techniques work by removing or altering one of the factors that the microbes need. Some examples are shown in the table.

Preservation method	Details of method	How decay is prevented
canning	food is heated in a can to about 100 °C and then the can is sealed	the high temperature kills the microorganisms; water and oxygen cannot get into the can after it is sealed
cooling	food is kept in refrigerators at about 5 °C	the low temperature slows down the growth and respiration of microorganisms
drying	dry air is passed over the food, sometimes, in a partial vacuum	microorganisms cannot respire or reproduce
freezing	food is kept in a freezer at about −18 °C.	microorganisms cannot respire or reproduce because their chemical reactions are slowed down
adding salt or sugar	food is stored exposed to a high sugar or salt concentration.	the sugar or salt draws water out of the microorganisms
adding vinegar	the food is soaked in vinegar	the vinegar is too acidic for the microorganisms preventing their enzymes from working

▲ Pickling food helps preserve it

c Why do you think that food still goes bad in a refrigerator?

d Suggest why salt or sugar draws water out of the microorganisms.

Some of these food preservation techniques have been used for thousands of years. This was particularly important in hot countries where decay would happen rapidly.

Food can also be preserved by adding artificial chemicals or additives to the food. These chemicals are not popular with everybody. Some people say that they add unpleasant flavours to the food. Others claim that they have side effects on the body.

Keywords

decay • detritivore • detritus • food preservation • saprophyte • saprophytic nutrition

Astronaut ice cream

It is now possible to buy packets of ice cream that do not need to be kept in the freezer. This ice cream tastes just like ice cream except it is not cold.

It was developed for astronauts to take into space and uses the process of freeze-drying. The ice cream is made in the normal way by freezing the ingredients in order to change the water into ice. The machine then pumps out the air, creating a vacuum. This lowers the pressure inside the ice cream. If the ice cream is then slightly warmed, the ice turns straight into water vapour without becoming liquid. This is called sublimation. The water vapour is removed, leaving dry astronaut ice cream!

Freeze-drying is useful for a number of products as well for as making ice cream. This is because the food does not spoil so easily once it has been treated. It is also used to preserve certain medicines. Another advantage is that freeze-dried food is lighter. Freeze-drying will not prevent the food spoiling indefinitely because there is still a small amount of water in the food.

Questions

1 How is sublimation different from melting?

2 Why is food lighter once it has been freeze-dried? Why is this an advantage?

3 Suggest why food such as freeze-dried coffee is packed in airtight containers.

4 Explain why freeze-dried food will still spoil eventually.

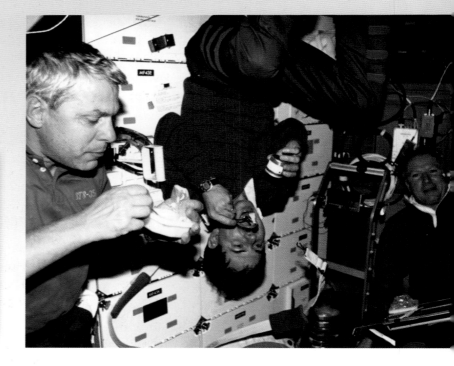

Cycles for life

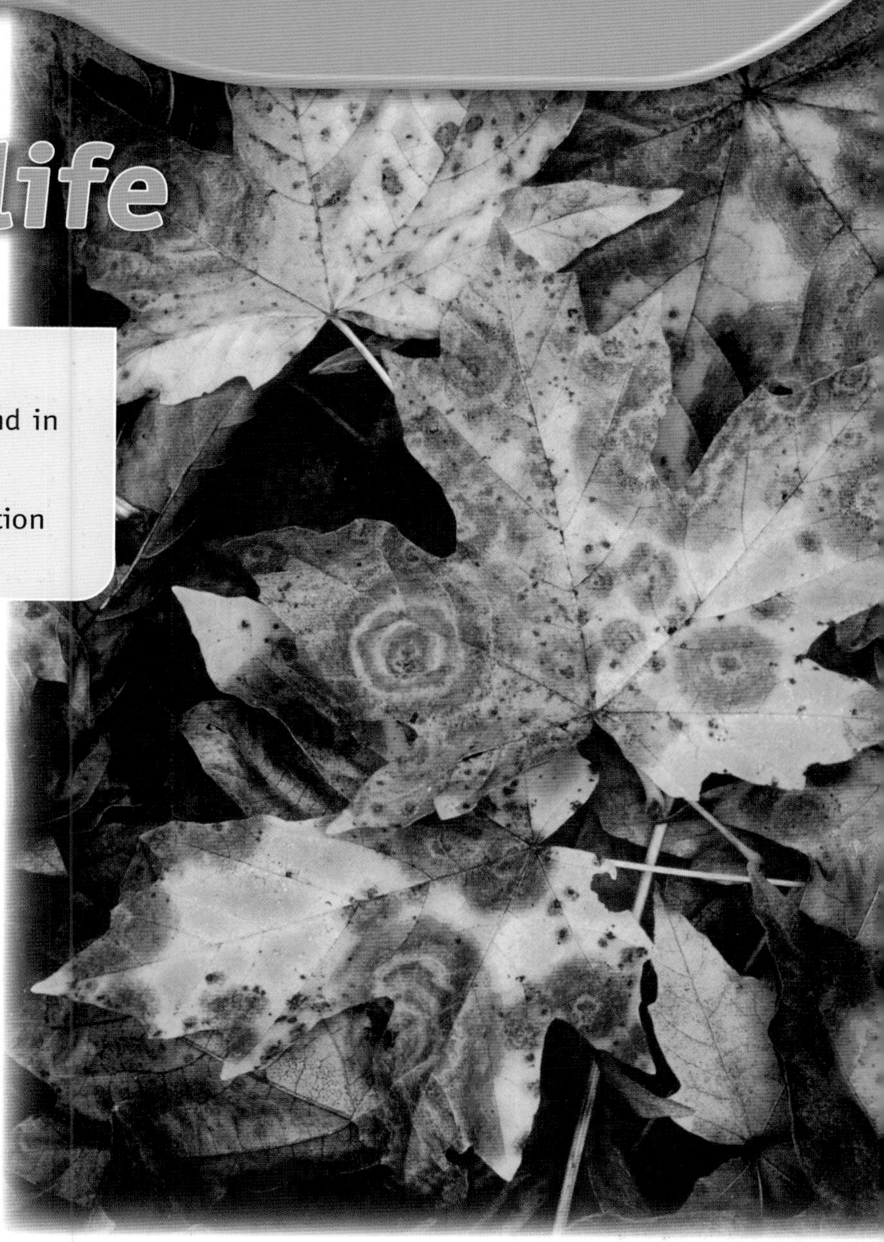

▲ *Recycling leaves*

In this item you will find out

- how carbon is recycled both on land and in the sea
- how nitrogen is recycled due to the action of bacteria

Recycling is becoming big business. Scientists have been getting very excited recently about a new development called nanotechnology. A nanometre is one thousand millionth of a metre and this is smaller than many atoms. Scientists now think that they can make structures about the size of 1–100 nanometres that can perform many roles. One important role that they are being designed for is to recycle minerals from many different materials.

a **Why do you think that scientists are under pressure to find ways of recycling materials?**

But humans are only beginners when it comes to recycling; bacteria and fungi have been in action for millions of years. Their job is to decompose dead plant and animal material and make all the chemical elements available again for living organisms. Without them, we would run out of carbon, nitrogen, oxygen and all the other elements that are needed for life.

Nitrogen often causes the largest problem for living organisms. Although we are surrounded by nitrogen it is not easy to use and it is desperately needed by plants. In 1909, the German scientist Fritz Haber developed a process to combine nitrogen from the air with hydrogen.

The ammonia that is produced can be turned into fertiliser for plants. Now about half of the nitrogen needed by all the plants grown in the world comes from the Haber Process.

However, providing all the energy needed for the Haber Process results in pollution, and so does the use of the fertilisers that are being produced.

Amazing fact

In every kilogram of soil there are about 5 g of living organisms many of which are microorganisms.

The carbon cycle

▲ *Graphite is a form of carbon*

The element carbon is the basis for all molecules that make up living organisms. Carbohydrates, proteins and fats all contain carbon. In nature, pure carbon is found as diamonds and graphite, but animals and plants cannot use this carbon.

The main source of carbon is carbon dioxide in the air but there is only one way that it can get into living organisms. This happens when plants photosynthesise.

This process traps the carbon inside carbon compounds and it is then passed from organism to organism along food chains or food webs. It returns to the air in carbon dioxide when plants and animals use the carbon compounds in respiration. This cycling of carbon is shown in the diagram.

The decomposers, bacteria and fungi in the soil, also release carbon dioxide when they use dead material for respiration.

Sometimes dead animals and plants do not decompose but over millions of years they are changed into fossil fuels. This process of fossilisation traps carbon in coal, oil and gas. Burning (combustion) of these fossil fuels releases this carbon again as carbon dioxide.

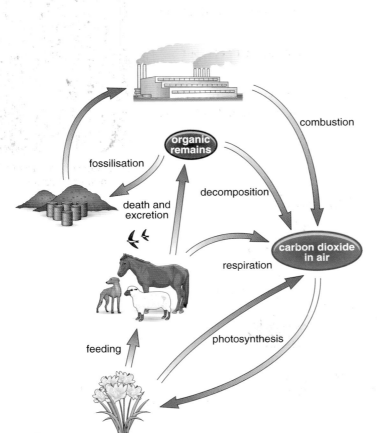

The carbon cycle ▶

Carbon in the sea

Carbon can also get locked up by organisms in the sea. Microscopic plants use carbon dioxide in photosynthesis and marine organisms use carbon to make shells. These shells are made of **carbonates**. When the organisms die the shells sink and get compressed at the bottom of the sea. They turn to limestone rock.

b **What do you think compresses the shells?**

Over the years this limestone rock can get worn away by weathering or more suddenly by volcanic activity. Carbon dioxide is released into the air and joins the cycle again.

The nitrogen cycle

Plants and animals are surrounded by air that contains 78% nitrogen but they cannot use it directly because it is too unreactive.

Plants take in nitrogen as nitrates through their roots and use the nitrates to make nitrogen compounds (proteins) for growth. This protein passes along the food chain or web as animals eat plants and other animals.

 c Explain why the quantity of nitrogen in the animals decreases along the food chain.

Eventually all this trapped nitrogen is released when decomposers break down nitrogen compounds in dead plants and animals.

The nitrogen cycle is more complicated than the carbon cycle because four different types of bacteria are involved instead of just one. This is shown in the diagram below.

▲ These cliffs are made from chalk – a form of limestone

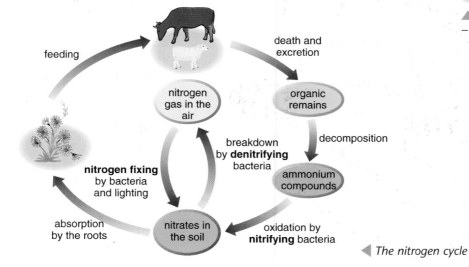

◀ The nitrogen cycle

Soil bacteria and fungi, acting as decomposers, convert proteins and **urea** into ammonia. This is poisonous to plants but **nitrifying bacteria** turn it into nitrates.

Denitrifying bacteria turn some of these nitrates into nitrogen gas. **Nitrogen-fixing bacteria** that live in the roots of plants of the pea family can make use of the nitrogen gas in the air and return it to the cycle. This is called fixing nitrogen. Lightning can also fix nitrogen.

 d Nitrogen is very unreactive. Why do you think nitrogen can react when there is a lightning strike?

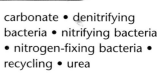

Keywords

carbonate • denitrifying bacteria • nitrifying bacteria • nitrogen-fixing bacteria • recycling • urea

Feeding fish and rice

A little red fern is proving to be a very valuable plant for the Chinese. *Azolla* is a tiny fern that floats on the surface of lakes and rivers. The important thing is that the fern contains bacteria that live inside the leaves. These bacteria can fix nitrogen and so provide the fern with nitrogen-containing chemicals.

The Chinese have started to put this partnership to work in their rice fields. Rice is grown in large flooded fields. The Chinese also farm fish in the same flooded fields. This provides them with extra food and the fish eat the rice pests.

The problem is that there is not much food for the fish in the fields. This has been solved by adding *Azolla*. The fish eat the fern, which is rich in amino acids, and grow quickly. The fish also help the rice to grow by producing nitrogen-rich faeces that act as a fertiliser.

The yields that can be achieved are shown in the table.

▲ Azolla *fern*

▼ *The bacteria that live in* Azolla *leaves*

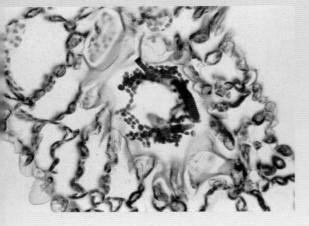

Organisms growing in fields	Yield of organism from the field (kg/ha)	
	Rice yield	**Fish yield**
rice only	6930	–
rice and *Azolla*	8085	–
rice and fish	7656	150
rice, fish and *Azolla*	9324	350

Questions

1 The bacteria in *Azolla* fix nitrogen for the fern. What does *Azolla* make using use these nitrogen-containing chemicals?

2 Suggest what the fern gives the bacteria in return.

3 Construct a food web to include rice, fish, *Azolla* and rice pests.

4 The nitrogen in the fish faeces becomes available to the rice. Explain how this can happen.

5 Work out the total yield of fish and rice per hectare when they are grown without Azolla and when grown with *Azolla*.

6 What is the percentage increase produced by using *Azolla*?

B4a

1 The diagram shows a section through a leaf.

Write down the letter on the diagram that matches the following structures.

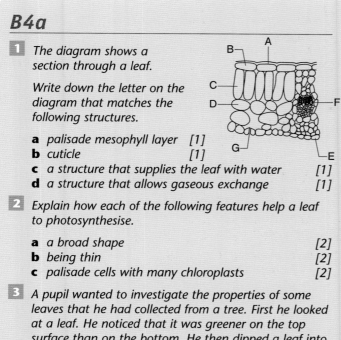

a palisade mesophyll layer [1]
b cuticle [1]
c a structure that supplies the leaf with water [1]
d a structure that allows gaseous exchange [1]

2 Explain how each of the following features help a leaf to photosynthesise.

a a broad shape [2]
b being thin [2]
c palisade cells with many chloroplasts [2]

3 A pupil wanted to investigate the properties of some leaves that he had collected from a tree. First he looked at a leaf. He noticed that it was greener on the top surface than on the bottom. He then dipped a leaf into hot water. He noticed that small air bubbles appeared on the bottom surface of the leaf but not on the top. He then wanted to look at the surface of the leaf but his teacher told him that he would need to make an impression of the surface rather than putting the leaf under the microscope.

a Use your knowledge of leaf structure to explain why leaves look greener on the top than on the bottom. [2]
b Where in the leaf does the air in the air bubbles come from? [1]
c Explain why there are only air bubbles on the bottom of the leaf. [2]
d Suggest why is it necessary to make an impression of the leaf to see details of the leaf surface. [2]

B4b

1 The diagram below shows two solutions in a glass beaker. They are separated by a partially permeable membrane. In compartment A there is a concentrated sucrose solution. In compartment B there is distilled water.

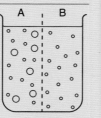

○ sucrose
○ water molecule

a Copy the diagram and draw an arrow to show which way the water will move. [1]
b What is the name of the process that causes the water to move? [1]
c The sugar molecules cannot move through the partially permeable membrane. Why is this? [1]

2 What is the difference between osmosis and diffusion? [2]

3 Explain why animal cells burst when placed in water but plant cells do not. [2]

B4c

1 Copy and complete the following sentences by entering a word or words in each of the lines.

In a plant stem the xylem and phloem are arranged in groups called __(1)__. In the stem the __(2)__ is on the outside and the __(3)__ is on the inside of each of these groups. The xylem transports __(4)__ up the stem and the phloem transports __(5)__ in __(6)__ directions. The evaporation of water out of the leaves is called __(7)__. [7]

2 The potometer shown on page 50 is used for another experiment. This time the leaves are not covered with grease but the experiment is carried out under different conditions. Some results are shown in the table.

Conditions	Distance water moved from start position in ten minutes/mm
normal lab conditions	30
with a kettle boiling in the room	25
with the kettle boiling and the lights off	20

a Explain why the water moves along the tube in the potometer. [2]
b Fully explain the results of this experiment. [5]
c Explain what would happen if the experiment was repeated with a fan pointed at the plant. [3]

3 Explain how the following tissues are adapted for the job that they do in a plant.

a phloem [1] **b** xylem [2]

B4d

1 Jane buys a packet of fertiliser to use on her garden. On the packet it says:

'This fertiliser contains all the minerals needed for your plant to grow. Simply dissolve in some water in a watering can and water the soil. Only use a small amount of the powder, plants are not used to too many minerals.'

a Why should water and a watering can be used to apply the fertiliser? [2]
b Why are plants not used to too many minerals? [1]
c Which mineral in the fertiliser would be needed for cell growth? [1]

2 An experiment was carried out to investigate the effect of growing plants with different minerals available. The results are shown in the photograph on page 54.

  a Suggest why each of the solutions in this experiment was made using distilled water and not tap water. [2]

  b Which plant grew best and why? [2]

  c What is the difference in colour between the nitrogen-free plant (–N) and the plant grown with all the necessary minerals (complete)? [1]

  d What effects does the lack of phosphate (–P) have on the plant? [2]

3 The following apparatus can be used to see how fast plants take up minerals.
The minerals are taken up by active transport. The graph shows how fast the minerals are taken up.

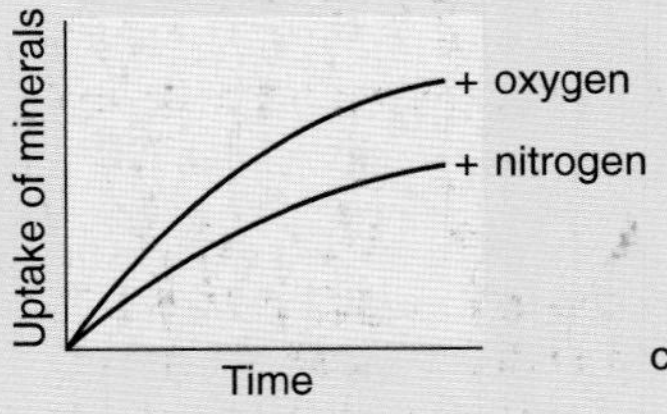

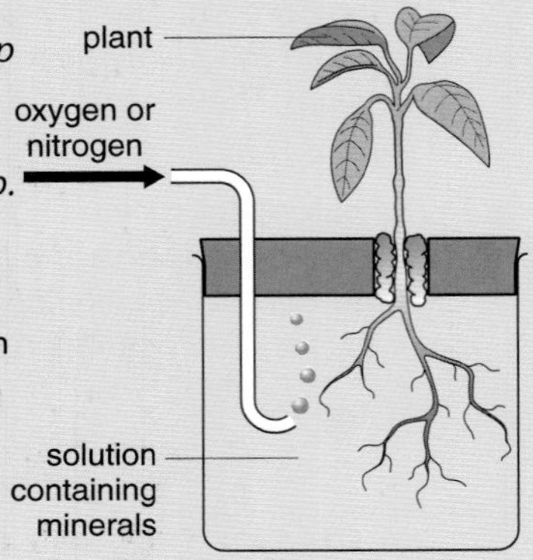

  a Why is active transport used to take up minerals? [1]

  b What is the difference in the result when nitrogen is bubbled into the solution rather than oxygen? [1]

  c Explain the difference in uptake when nitrogen is used rather than oxygen. [3]

B4e

1 Several animals live in a garden. The table shows what they eat.

Animal	Food
snail	grass
rabbit	grass
mice	grass seeds

Animal	Food
fox	rabbit, mice
blackbird	snail

  a Which organism in the table is a producer? [1]

  b Draw a food web for the organisms listed in the table. [3]

  c Many tiny fleas live in the fur of the fox. Draw a pyramid of numbers for this food chain: grass plants → rabbits → fox → fleas [2]

  d Draw a pyramid of biomass for the same food chain and explain why it looks different from the pyramid of numbers. [3]

2 The diagram shows energy being lost as it passes through a food chain.

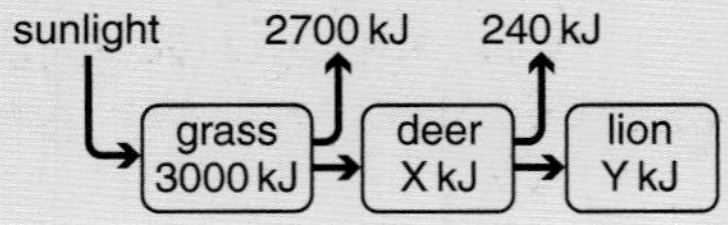

sunlight   2700 kJ   240 kJ

grass 3000 kJ → deer X kJ → lion Y kJ

  a Calculate the amount of energy available to the deer for growth. [1]

  b Calculate the amount of energy available to the lion. [1]

  c What percentage of the original energy from the plant reaches the lion? [2]

  d Name two ways in which energy is lost by the deer. [1]

  e Use the information on the diagram to explain why food chains rarely have more than five levels. [2]

3 The diagram shows a process for converting biomass into a fuel called gasohol.

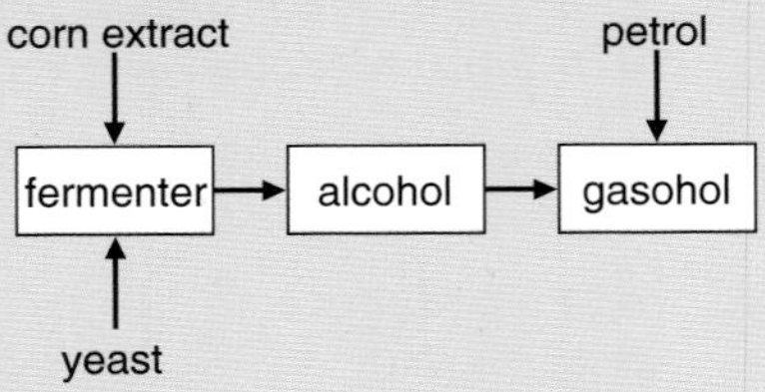

corn extract → [fermenter] → [alcohol] → [gasohol] ← petrol
↑ yeast

  a Corn extract is the biomass that is placed in the fermenter. Where has the energy that is trapped in the corn come from? [1]

  b Why is yeast added to the corn extract? [2]

  c This process is used in countries such as Brazil. These are hot countries with long hours of sunshine. They do, however, have limited supplies of petrol. Explain why the use of gasohol as a fuel is ideal in countries such as Brazil. [2]

B4f

1 Tomato plants can be grown in glasshouses. They are often grown without soil.

  a Why are tomatoes often grown in glasshouses? [1]

  b What name is given to the method of growing tomatoes without soil? [1]

  c Explain why these two methods mean that tomatoes can be grown in areas where they could not otherwise have been grown. [2]

2 A small red spider often feeds on the leaves of tomato plants. It is possible to control the red spider by releasing another type of spider called Phytoseiulus into the glasshouse.

  a Suggest how Phytoseiulus controls the red spider. [1]

  b What is the name given to this type of control? [1]

  c Why is this type of control ideal for a greenhouse but less easy to use in a field? [1]

  d Why must the owner of the greenhouse stop using pesticides when he uses Phytoseiulus? [1]

3 An organic farmer may use a range of different methods to try to increase his yield. Explain what is meant by the following methods and how they can increase yields.

  a crop rotation [3]   **b** manuring the soil [3]

4 Intensive farmers use a range of different methods compared to organic farmers.

    a Suggest why battery farming can produce greater yields of meat from animals. [2]

    b Why are some people against battery farming? [2]

B4g

1 Below are types of food preservation techniques and explanations of how they work. Link each technique with the explanation of how it works.

1 freezing	a microorganisms are killed by heat
2 adding vinegar	b temperature is too cold
3 adding salt	c pH is too low
4 canning	d microorganisms are dried out [2]

2 Different organisms are responsible for decomposing dead leaves. These are:

- earthworms, which may be about 5 mm in diameter
- small insects, such as maggots and woodlice, which may be 2–4 mm wide
- microorganisms which are smaller than 0.005 mm wide.

A scientist decided to investigate how fast leaves decompose. He put leaves into three different bags and buried them in the soil. Each bag was made of nylon with different sized holes. Every two months he dug up the bags and measured how much of the leaves had disappeared.

Here are his results.

	Disappearance of the leaves/%		
Month	**Bag with 7 mm**	**Bag with 4 mm**	**Bag with 0.005 mm**
June	0	0	0
August	25	8	0
October	70	20	2
December	75	25	3

    a Which of the three types of organisms can get into each of the bags? [1]

    b In which bag do the leaves decay the fastest? [1]

    c Explain why the leaves decay at different rates in the three different bags. [4]

    d How does the rate of decay change in November and December compared to July to October? [1]

    e Explain this difference in rate. [2]

3 Many bacteria and fungi are saprophytes.

    a What does a saprophyte feed on? [1]

    b How does a saprophyte digest its food? [2]

B4h

1 The diagram shows part of the carbon cycle.

    a Write down the name of the process represented by each of the letters on the diagram.
Choose your processes from this list. You can use each process once, more than once or not at all.

decomposing   eating   fossilising photosynthesising   respiring [5]

The carbon may get trapped in fossil fuels.

    b How does this happen? [2]

    c How can this carbon rejoin the carbon cycle? [1]

2 Read the following passage about nitrogen and answer the questions that follow.

Nitrogen is an important element for all organisms from grass to giraffes and pansies to pigs. All these organisms use nitrogen for growth. But they all have a problem. Although they are surrounded by plenty of nitrogen in the air, they cannot use it very easily. Animals rely on plants to get nitrogen from the soil. The plants absorb this nitrogen combined in minerals. The animals can then eat the plants!

    a What do animals and plants make from nitrogen that is so important for growth? [1]

    b The article says that there is plenty of nitrogen in the air. What is the percentage? [1]

    c Why is it so difficult for animals and plants to use this nitrogen? [1]

    d What is the main mineral taken up by plants that contains nitrogen? [1]

3 The diagram shows part of the nitrogen cycle.

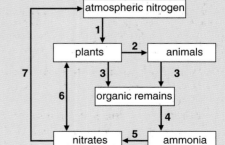

    a Write down the type of bacteria that carry out each of the processes 1, 4 and 5. [3]

    b The bacteria carrying out process 4 are converting chemicals in organic remains into ammonia. Write down the name of one of these chemicals. [1]

    c Explain how lightning can play a role in the nitrogen cycle. [2]

Gran has just had a hip replacement operation. She had lots of pain and could hardly walk. After the operation she was walking again after just a few days.

My father had a scare. Doctors found a blocked coronary artery but they were able to repair it. He came home after two days. He's taken up jogging now.

Every day there is news of a new breakthrough in medicine. Do you think that, one day, everyone will be able to live to be 100 years old?

- No one can be sure what the quality and length of their life will be, but everyone can try to follow a healthier lifestyle. What do we do when things go wrong? Scientists can extend and improve life by using a 'spare-part surgery' repair kit. Damaged organs are replaced by transplants or implants. One day we may be able to grow a supply of spare parts using human stem cells.

- Many people want to become parents. Because of genetic disorders or problems with infertility this is sometimes not possible. Advances in our understanding of genetics and reproduction have made it possible for many of these people to become parents.

- These treatments raise a number of economic, religious and ethical issues. Who should have the treatments? Is it right that scientific discoveries should be applied in these ways?

What you need to know

- Lifestyle plays an important part in your development and general health. A poor diet, not enough exercise and abusing drugs can reduce the quality and length of your life.

- Advances in medicine make it possible to repair or replace some damaged or diseased organs.

- There are treatments that help some infertile couples or people with genetic disorders to have normal, healthy babies.

Bare bones

In this item you will find out

- about the advantages of an internal skeleton

- how bones grow and repair themselves and how damaged joints can be replaced

- how bones and muscles work together in joints to produce movement

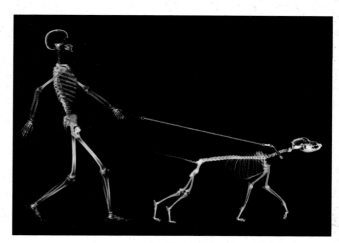

▲ The skeleton is a flexible framework that allows movement and protects your internal organs

Imagine what your life would be like without a skeleton. Your skeleton acts as a safety cage that protects your internal organs. The skull protects your brain, and your spine protects the spinal cord. Moving a person with spinal injuries runs the risk of permanent damage to the nerves and may result in paralysis or death. The rib cage protects the heart and lungs and is also responsible for the breathing movements that inflate and deflate your lungs.

As you grow, your skeleton becomes stronger and larger. If your bones are damaged, they can repair themselves. Your skeleton is also a system of joints and levers that allows you to run and jump and ride bicycles. These joints can move in different ways to make you flexible and strong. It is important that your skeleton is correctly proportioned. The bones and joints in your legs are much longer and stronger than those in your arms. This enables them to withstand greater forces.

Humans have internal skeletons called endoskeletons. Some animals, such as insects and spiders, have external skeletons called exoskeletons.

Internal skeletons are generally smaller, lighter and allow a greater range of movement than external skeletons. They do not restrict growth so much. Animals with exoskeletons, like crabs, are heavy and clumsy. They must shed their skeleton to grow, making them more vulnerable to predators. The human skeleton has over 200 separate bones and five different kinds of joint. This variety of joints allows a wide range of movements to be made. If you watch a gymnast in action, the advantages of an internal skeleton become obvious.

Amazing fact

The largest skeleton of all is visible from space! It is the Great Barrier Reef formed from a skeleton of coral.

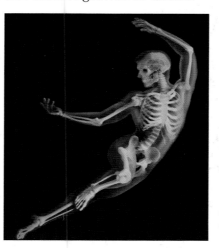

◀▲ Two types of skeleton. Which is the most successful?

a Explain three advantages of an internal skeleton compared to an external skeleton.

Growing bones

Your skeleton starts as **cartilage**. As you grow, the cartilage is slowly replaced by calcium salts and phosphates that turn it into bone. This is called **ossification**. This is why a diet with lots of calcium, phosphorus and vitamin D is important for strong bones. While you are an adolescent, the cartilage near the ends of your limb bones produces more bone. This called a growth spurt. Your sex hormones help to control the production of bone.

Eventually most of the cartilage is turned into bone and the skeleton stops growing. Adults have cartilage in the nose, at the ends of the ribs and around the surfaces of moveable joints. Cartilage and bone are both living tissues. They can suffer from infections but they can grow and repair themselves.

What's inside a bone?

The diagram shows the structure of a long bone. It consists of a head and a long shaft. The head is covered in cartilage while the shaft contains bone marrow with lots of blood vessels in it. Long bones are hollow and are much stronger than a solid structure of the same mass.

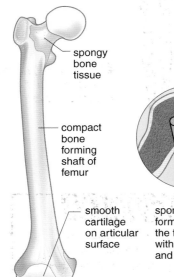

- spongy bone tissue
- compact bone forming shaft of femur
- smooth cartilage on articular surface

▲ *The femur is long and hollow making it strong but light*

spongy bone forms the head of the femur and withstands jolts and shocks

compact bone is important for strength and rigidity

cartilage provides a smooth hard, slippery, load-bearing surface

Damaging bones

Have you ever broken a bone? Bones can easily be damaged in accidents. Although they are strong they can be broken by a sharp knock. A break involving just bone is a simple fracture. Compound fractures involve other tissues, and the broken bones may stick out from the skin. It is dangerous to move someone if you think they have a fracture. If their spine gets damaged then they could be paralysed or even die.

Elderly people often suffer from **osteoporosis**. This is due to the loss of minerals from the bone. Osteoporosis leads to a gradual softening of the bones and is caused by a poor diet or lack of exercise. The National Health Service now offers a simple test in which the density of a person's heel bone is measured using ultrasound. This enables the doctor to spot the early signs of osteoporosis and to prescribe a course of treatment to prevent the bones weakening further. If someone suffers from osteoporosis they are more likely to suffer a fracture if they fall.

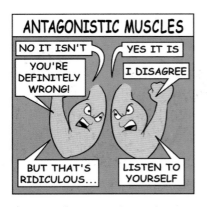

ANTAGONISTIC MUSCLES

NO IT ISN'T

YES IT IS

YOU'RE DEFINITELY WRONG!

I DISAGREE

BUT THAT'S RIDICULOUS...

LISTEN TO YOURSELF

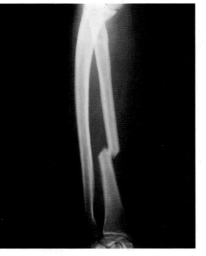

▲ *Broken bones must be held firmly in position while they repair themselves*

b What are the symptoms of osteoporosis?

c Suggest two ways in which osteoporosis might be prevented.

The human machine

Your skeleton consists of 206 bones which are held together by a number of different kinds of joint.

- The bones of the skull form fixed joints that fit together to make the cranium and protect the brain.
- **Synovial joints** are moveable joints that allow different degrees of movement. In a synovial joint the ends of the bones are covered with smooth cartilage which reduces friction. The moving surfaces are lubricated by synovial fluid. This fluid is secreted by the synovial membranes which also act like an 'oil seal'.
- Your limbs are joined to your body at the shoulder and hip by ball-and-socket joints which allow circular movements in more than one plane.
- Your elbows and knees are examples of hinge joints and move in one plane. **Ligaments** keep the hinge joint together while tendons transmit the pull of muscles to the bones.
- Your wrists and ankles contain lots of small bones that form sliding or gliding joints.
- You also have joints between your vertebrae, fingers, toes and at the ends of the ribs to which muscles are attached. These allow different ranges of movement.

 d Ligaments and tendons are made of tissue that is very strong and flexible but is not very elastic. Suggest why these are useful properties for ligaments and tendons to have.

 e Explain how a synovial joint allows easy movement.

Flexing your muscles

Muscles can only exert a force when they contract, so they work in pairs. When you bend your arm the **biceps** muscle contracts and the **triceps** muscle relaxes. This action is called flexion. When you straighten your arm, the triceps contracts and the biceps relaxes. The biceps and triceps are called **antagonistic muscles** because they work in opposite ways. Your arm bending and straightening is an example of a lever.

cranium (fixed joint)
scapula
shoulder joint (ball-and-socket joint)
biceps
humerus
triceps
elbow (hinge joint)
tendons
ulna
radius
wrist joint (sliding joint)

◀ *The skeleton provides a framework for the muscles to move*

Keywords

antagonistic muscles
• biceps • cartilage •
ligament • ossification •
osteoporosis • synovial
joint • triceps

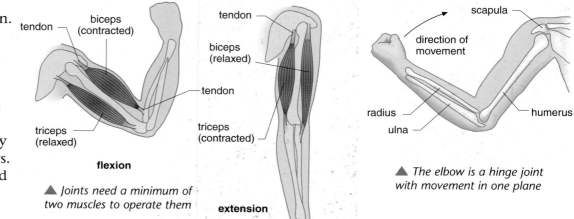

tendon
biceps (contracted)
triceps (relaxed)

flexion

▲ *Joints need a minimum of two muscles to operate them*

tendon
biceps (relaxed)
tendon
triceps (contracted)

extension

scapula
direction of movement
radius
ulna
humerus

▲ *The elbow is a hinge joint with movement in one plane*

New joints for old

Morag suffers from arthritis. Her hip joint has lost cartilage from the moving parts of the joint. Extra bone has grown and deformed the shape of the joint. This means that her movement is restricted. Her hip has become inflamed, which is very painful. Her doctor says that her hip joint must be replaced with an artificial one made of strong metal alloys and tough plastic.

Morag's knee is also giving her trouble. It can be replaced by an artificial joint as well, but it is not as bad as her hip.

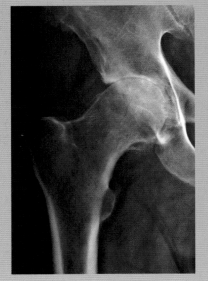

▲ Arthritis causes pain and limits movement by damaging joints

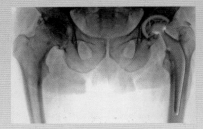

▲ Replacement joints enable people to return to an active and pain-free life

▲ A replacement hip joint

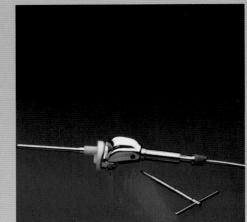

▲ A replacement knee joint

During Morag's hip replacement, the surgeon will cut away the damaged bone. The replacement joint will be fixed into the hip and thigh using 'liquid bone' made from treated, powdered coral. Adding acid to the coral paste makes it set like cement within a few minutes. Within hours it will be as hard as real bone, and Morag will be able to start walking again in a few days.

Replacement joints make it possible for people to regain their mobility and be free from pain. Not only does this enable a patient to live a more active life, it also allows them to be more independent. It can improve other aspects of their mental and physical health. Improvements in surgery and the use of modern materials mean that people recover quicker and the joints last longer. Accidental injuries can be repaired and amputations can sometimes be avoided. Artificial joints are not rejected and they can be adjusted to fit the patient. However, it is not possible to repair joints where the bones are weakened by disease and some people may not be fit enough to withstand the surgery.

Questions

1 Damaged joints can be replaced by artificial joints made of metal and plastic. What are the advantages of using artificial joints rather than using transplants?

2 What are the symptoms of arthritis?

3 Suggest why it is not possible to repair every damaged joint with an artificial one.

The beat of life

In this item you will find out

- about single and double circulatory systems
- how the heart works
- how heart rate is measured and controlled

▲ *A human heart pumps about 10 tonnes of blood per day*

The heart is one of your most important organs. It beats constantly from before birth until we die. Scientists throughout the centuries have been fascinated by how it works, and how to fix it when it goes wrong.

The heart is part of the circulatory system. Fish have a **single circulatory system**. Their two-chambered heart pumps deoxygenated blood to the gills. Oxygenated blood flows to the body organs and back to the heart in one continuous circuit.

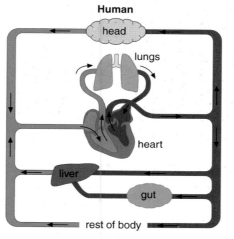

Double circulation in humans

Key

■	Oxygenated blood
■	Deoxygenated blood

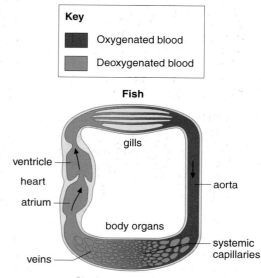

Single circulation in fish

▲ *Different kinds of circulation*

Humans have a four-chambered heart powering a **double circulatory system**. In one circuit, deoxygenated blood is pumped from the heart to the lungs. Oxygenated blood returns to the heart. In another circuit, oxygenated blood is pumped to the body. Deoxygenated blood returns to the heart and the process is repeated.

a In which of the two circuits in humans would the blood be under most pressure? Suggest why.

The volume of blood flowing through an organ varies. During exercise, the heart beats harder and faster. Small arteries supplying blood to the organs control its flow by becoming wider or narrower. The table shows how blood flow changes during exercise.

b Using the table: (i) Suggest two reasons why blood flow to the heart changes during exercise. (ii) Which organs have a reduced blood flow during exercise? How is this achieved? (iii) Which structure shows the greatest change in blood flow during exercise?

Structure	Blood flow (cm³/min)	
	at rest	during exercise
Brain	700	740
Heart	200	750
Kidneys	1000	600
Liver	1400	600
Lungs	100	200
Skeletal muscles	750	12 000
Skin	300	1900

Finding out about circulation

Galen (c130–200 AD) studied anatomy in Rome. He thought blood was made in the liver and that it flowed through the heart to the body and back again. He described the circulation as a tidal movement.

Galen's ideas were believed until a doctor, William Harvey (1578–1657), proved they were false. Harvey calculated that the heart pumped 100 times the body's blood volume every hour, so the same blood must circulate round the body. He showed that veins contain valves so blood must travel one way to the heart. He found that arteries, if compressed, filled up with blood on the side closest to the heart.

Harvey also predicted the existence of capillaries between arteries and veins even though the microscopes needed to see them had not been invented. Harvey's scientific approach disproved ideas that had been believed for nearly 1500 years!

The living pump

The human heart is a double pump with four chambers, two atria and two ventricles. Both sides of the heart pump at the same time. Valves between each atrium and each ventricle stop blood flowing backwards.

The right atrium fills with deoxygenated blood from the vena cava. Blood then passes into the right ventricle. When the right ventricle contracts it pumps deoxygenated blood along the pulmonary artery to the lungs. Oxygenated blood returns along the pulmonary vein to the left atrium. Blood enters the left ventricle which pumps it along the aorta. This is called the **cardiac cycle**.

 The left ventricle has to pump blood hard enough to travel round the body. Why do you think that it has thicker walls than the right ventricle?

d Suggest why it is important that the heart does not pump blood backwards.

e Use the table on page 153 to explain why arteries need thick, muscular walls and why veins have thinner walls.

 Arteries branch into millions of capillaries as they enter organs. Oxygen and glucose diffuse from these capillaries to the organs. Why do you think capillaries have thin, permeable walls and why do you think there are so many of them?

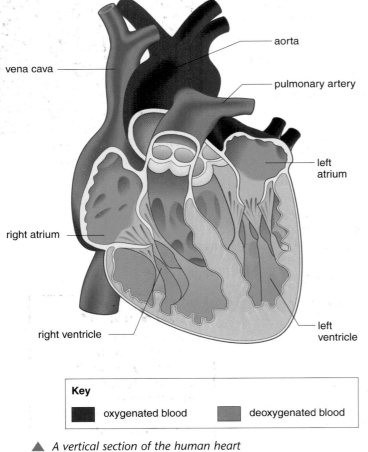

aorta

vena cava

pulmonary artery

left atrium

right atrium

right ventricle

left ventricle

Key

▪ oxygenated blood   ▪ deoxygenated blood

▲ *A vertical section of the human heart*

	Artery	Vein	Capillary
Diameter	2–25 mm	2–30 mm	Less than 0.1 mm
Type of wall	Thick, muscular	Thin, less muscular	Thin, permeable
Valves	No	Yes	No
Pressure	High (10–16 kPa)	Low (1 kPa)	Medium (2.5 kPa)

Your beating heart

Your heart muscle contractions are controlled by a **pacemaker**. The pacemaker produces nerve impulses. It consists of two groups of cells, the **sino-atrial node (SAN)** and the **atrio-ventricular node (AVN)**. In the SAN, nerve impulses start the heartbeat by making the right and left atria contract at the same time. This fills the ventricles with blood. The nerve impulses then reach the AVN and both ventricles contract. This then forces blood out of the heart.

An **electrocardiograph (ECG)** can detect the electrical current produced by the pacemaker cells. The signals produced during each heartbeat are detected by electrodes placed on a person's chest. They can then be displayed on a TV screen or recorded on a paper printout.

An **echocardiogram** bounces ultrasound waves off the heart to show it beating. It can produce live, moving pictures of the inside of the heart as it pumps blood.

When you exercise your muscles need more blood and so the heart is made to beat harder and faster. If you are stressed or scared your heart rate also increases. A hormone called **adrenaline** stimulates your heart and redirects blood away from your skin and digestive system to your muscles.

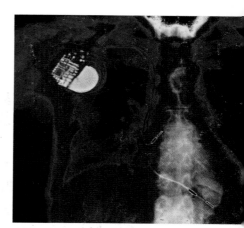

▲ A pacemaker can be used to keep the heart beating properly

g Why do you think your muscles need more blood when you exercise?

h Explain why people go pale and feel 'butterflies' in their stomach when they are frightened.

Controlling your heartbeat

If you have a problem with your heartbeat it may need to be controlled artificially. You can have a miniature artificial pacemaker implanted under your skin. It works by responding to the amount of activity you are doing and stimulating your heart to beat to match this activity.

Keywords

adrenaline • atrio-ventricular node (AVN) • cardiac cycle • double circulatory system • echocardiogram • electrocardiograph (ECG) • pacemaker • single circulatory system • sino-atrial node (SAN)

Casualty!

▲ *A busy casualty department*

The casualty department in a hospital is a very busy place. Patients with suspected heart problems are given an ECG test. Electrodes stuck to the patient's chest detect the weak electrical impulses produced each time the heart beats. These impulses are amplified, displayed on a TV screen, and printed as a graph.

The doctor can see disturbances in the heart's normal rhythm or tell if the person has had a recent heart attack. An ECG can pinpoint dead tissue in the heart muscle and detect whether the heart is enlarged or working under strain.

- Mr Patel is 53. He is a heavy smoker and is overweight. He has been experiencing tight chest pains.
- Jane Green has collapsed. She has been weak and short of breath for some time. She feels a 'fluttering' in her chest and has a cold sweat.

The doctors at the hospital use the ECGs to make a diagnosis for each patient. They know the following:

- Tachycardia is a disturbance in the rhythm of the heart. The heart beats faster and the QRS waves have variable heights.
- During a heart attack, ECG traces often show upside-down T waves.

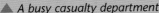

Key

P	Impulse causing atria to contract
QRS	Impulse causing ventricles to contract
T	Recovery before next heart beat

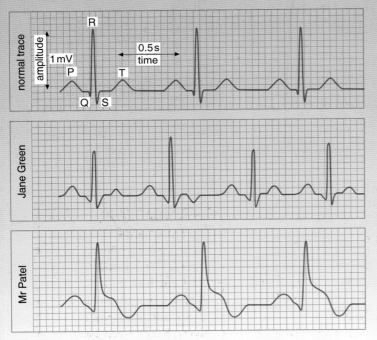

▲ *The ECG traces of two people*

Questions

Compare the ECG traces of Mr Patel and Jane Green with the normal trace.

1 For each patient, state:
  (a) how their trace differs from the normal one
  (b) what you think is wrong with the patient's heart. Give reasons for your answer.

2 Suggest why ECG machines contain amplifiers.

Your heart in their hands

On 3 December 1967 a surgeon in South Africa, Dr Christiaan Barnard, made headline news by announcing the world's first successful **heart transplant**. He proved that it was possible to replace a diseased heart with a healthy one taken from a dead donor. Other surgeons developed the operation with increased success.

There have been several attempts to make artificial hearts. Some of these are **heart assist devices** which do the job of a failing heart until a suitable donated organ can be found.

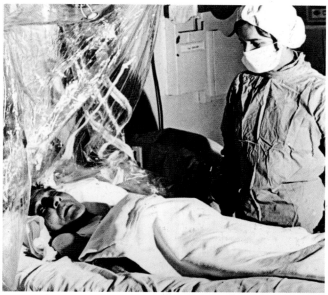

▲ *Louis Washkanski, the world's first successful heart transplant patient*

Artificial hearts must be reliable, but the main problem is finding a suitable power source. Some use rechargeable batteries and others use compressed air, but both limit the patient's activities.

There is a shortage of suitable donated hearts. Hearts must be healthy and match the patient's size and tissue type. The patient must constantly take anti-rejection drugs that lower their resistance to disease. However, many recipients return to work and go on to enjoy many years of good health.

Genetics, lifestyle and infection can all contribute to heart disease. The message for you is clear. Eat sensibly; reducing your fat and salt intake, and take regular exercise. This lowers your blood pressure and helps maintain your ideal weight. Setting aside some time for a hobby reduces stress. People who limit their intake of alcohol and stop smoking will improve their heart and circulatory system.

Amazing fact

By the year 2000 over 55 000 people had received heart transplants. In 2004, the world's longest-living heart transplant patient was still well 23 years later.

   i   Explain the difference between a heart transplant and a heart assist device.

   ii   Give two disadvantages of using heart assist devices.

   iii   Write down two things which must be checked when selecting a heart for a heart transplant.

A hole in the heart

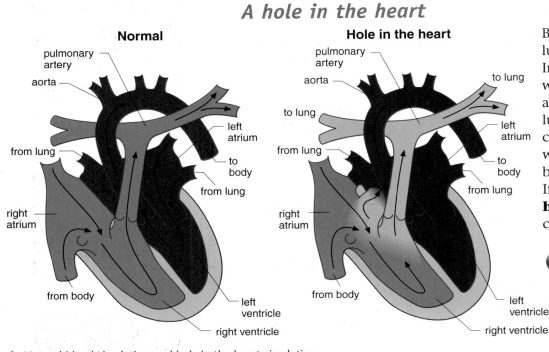

Normal

pulmonary artery
aorta
from lung
right atrium
from body
left atrium
to body
from lung
left ventricle
right ventricle

Hole in the heart

pulmonary artery
aorta
to lung
to lung
from lung
right atrium
left atrium
to body
from lung
from body
left ventricle
right ventricle

▲ *Normal blood circulation and hole in the heart circulation*

Before birth, the job of the lungs is done by the placenta. In a fetal heart, a hole in the wall between the two atria allows blood to by-pass the lungs so it flows in a single circulation. This hole closes when the baby is born so the baby has a double circulation. If this does not happen, the **hole in the heart** must be closed surgically.

b In a hole in the heart, blood can pass directly from the right atrium to the left atrium. Which organs are by-passed when this happens?

Faulty valves and blocked arteries

Sometimes valves inside the heart fail to develop fully or they may become damaged. Faulty valves can be repaired or replaced. Replacement valves may come from human donors or from genetically engineered pigs. There are also different types of artificial valve which can be used.

c Suggest one advantage and one disadvantage of using artificial valves to replace faulty heart valves.

Coronary heart disease occurs when blood flow to the heart muscle is reduced by fat deposits narrowing the coronary arteries. A heart attack occurs when a severe blockage prevents oxygen reaching some of the heart muscle cells and they die.

Bypass surgery may be used to improve the blood supply to heart muscle. Blood vessels from the leg are grafted around the narrowed sections of the artery to restore blood flow.

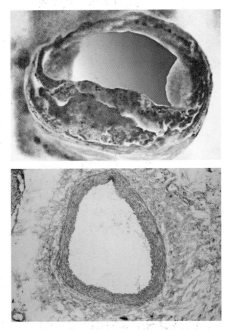

▲ *A diseased artery showing the build up of fat and a normal artery*

Life's blood

The structure and functions of the components of blood are shown below.

- **Plasma** (liquid part of blood): Plasma contains the products of digestion, dissolved oxygen, salts, waste products (urea, CO_2), hormones, clotting factors and antibodies.
- Red blood cells: Red blood cells contain haemoglobin, a protein which joins with oxygen to form oxyhaemoglobin in the lungs. It transports oxygen to the tissues and releases it when needed.
- White blood cells: Phagocytes change shape to swallow and digest bacteria. Lymphocytes include B cells that make antibodies in response to

'foreign' substances, T 'killer' cells that attack bacteria and memory cells that give immunity.

- Platelets: Platelets release chemicals which, together with proteins in the plasma, are needed for blood clotting. The plasma becomes sticky and turns into a jelly. This forms a clot that prevents blood loss.

 A red blood cell is a biconcave disc surrounded by a thin, flexible, semi-permeable membrane. It is packed with haemoglobin. Explain how the structure of a red blood cell suits its function.

 Explain how blood clots are formed.

Haemophilia is an inherited genetic disorder that can result in sufferers bleeding to death. Haemophiliacs' plasma lacks clotting factors which, together with platelets, are necessary for blood to clot.

The risk of blood clots causing a blockage can be reduced by using 'blood thinning' drugs such as **aspirin**, **warfarin** or **heparin**.

Other chemicals can also affect how the blood clots. Vitamin K, alcohol, green vegetables and cranberries all affect how the blood clots.

Blood transfusions

The surfaces of red blood cells carry antigens and plasma carries antibodies. Antibodies and antigens are sometimes called **agglutinins**.

People are divided into four blood groups, A, B, AB and O, depending on their antigens. The table shows the four types.

Antibodies in the plasma can cause clotting by reacting with antigens on the red blood cells. If blood group A red cells are **transfused** to a person with blood group B, they form clumps (agglutination). The A antigens on the red blood cells react with the anti a antibodies in the blood group B plasma.

 A and B antigen molecules have different shapes. Antibodies recognise different antigens by their shape. Suggest why blood transfusions sometimes go wrong.

Blood group	Antigens on red blood cells	Antibodies in the blood serum
A	A	anti b
B	B	anti a
AB	A and B	Nil (neither anti b nor anti a)
O	Nil (neither A nor B)	Both (anti b and anti a)

Because red blood cells from group O (the universal donor) have no antigens on them, they can be given to any group. An AB patient (the universal receiver) can receive blood cells from any **donor** because their plasma lacks both anti a and anti b antibodies.

 It is possible to transfuse red blood cells from a person with blood group A to a person with blood group AB but not the reverse. Explain why.

▲ A red blood cell

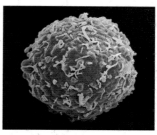

▲ A lymphocyte

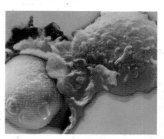

▲ A phagocyte engulfing a bacterium

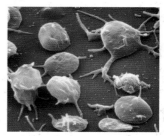

▲ Platelets

Keywords

agglutinins • aspirin • blood donor • blood plasma • blood transfusion • bypass surgery • haemophilia • heart assist device • heart transplant • heparin • hole in the heart • warfarin

Donors and dilemmas

▲ *Anita gives blood twice a year*

Anita gives blood regularly and her blood is used in blood transfusions. The discovery of blood groups has made transfusions safer. Transfusions have saved the lives of millions of people, including unborn babies. When Anita gives blood, her health is checked and her blood is processed to avoid the spread of disease.

Anita also carries an organ donor card. Many people put their details on a national computer database and carry organ donor cards, but there is still a shortage of suitable organs, as the graph for people waiting for a transplant shows. One way of increasing the supply of organs would be to assume that everyone's organs can be used unless the person has refused permission.

People have different views about organ transplants and blood transfusions. Advances in transplant surgery, biochemistry and medicine have made it possible to extend life when things go wrong, but this treatment is expensive. Some people think that the money spent on transplant surgery could be used to prevent and treat other diseases.

Scientists have discovered that we have different tissue types. This has made organ transplantation safer as organs are more closely matched. Anti-rejection treatments have improved but are expensive. Many different tissues and organs can now be transplanted or repaired with a high chance of success and expectations of long-term survival.

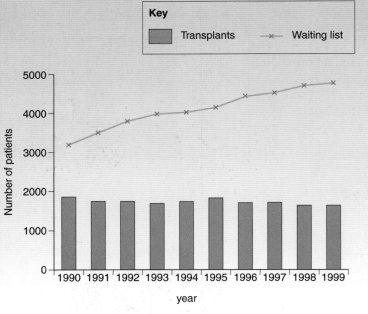

Key
- ▬ Transplants
- ─×─ Waiting list

▲ *The graph shows the number of kidney transplants and the size of the waiting list between 1990 and 1999*

Questions

1 What are the advantages and disadvantages of transplant surgery?

2 Look at the graph.
   (a) Describe how the number of patients waiting for kidney transplants changed between 1990 and 1998.
   (b) What was the average number of kidney transplants carried out each year during this period?
   (c) Suggest why the number of transplants did not increase significantly in this period.

3 A person may wish to donate their organs after death but has not mentioned this to their family. What problems might this cause?

4 Suggest why keeping details of donors on a computer database might be better than people carrying donor cards.

5 Some people think that all people should be organ donors unless they have opted out. Suggest reasons why people might agree or disagree with this idea.

Taking the air

In this item you will find out

- how fish, amphibians and humans exchange gases and the features of a gaseous exchange system

- about the ventilation of human lungs

- about some diseases of the lungs and respiratory system

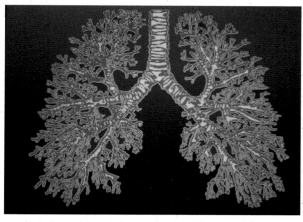

▲ *The human lungs take in over 300 million litres of air each year; enough to fill over four hot-air balloons*

All animals need to produce energy. To do this they have to respire. Aerobic respiration needs a plentiful supply of oxygen and is the most efficient way of producing energy. Fish use **gills** to extract dissolved oxygen from water. Frogs and other amphibians have gills in their tadpole stages so, like fish, must live in water. Before they can move on to land, amphibians must develop lungs. However, their lungs are relatively small and poorly developed so they use their skin to obtain extra oxygen. To survive, a frog's skin must always be moist.

 a **What environmental conditions does a frog need in order to survive?**

So how do fish carry out gaseous exchange under water? Fish use gills to get oxygen from water. Water enters the mouth, passes over the gill filaments and leaves through the gill flap. If you watch a goldfish you will see the gill flap pumping water over the gills.

The gills have three parts:

- gill rakers that filter out objects that could damage the gills
- the gill bar that supports the gill filaments
- feathery gill filaments containing blood capillaries.

As water travels between the filaments, oxygen diffuses into the capillaries and carbon dioxide passes into the water.

The gill flap can pump water but not air. If a fish is removed from water, the gill filaments stick together and the fish suffocates.

b **Why are the gill filaments less efficient when the fish is in air even though air contains oxygen?**

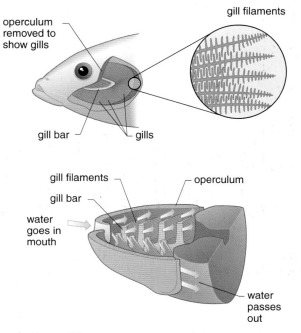

▲ *How a fish exchanges gases with its surroundings*

Exchanging gases

In our human gaseous exchange system we take in oxygen and give out carbon dioxide. Our lungs are constructed so that that oxygen and carbon dioxide are exchanged efficiently:

- the alveoli have a large surface area, usually in a compact space
- there are thin, moist, permeable membranes so gases can diffuse through them easily
- there is a plentiful blood supply to unload carbon dioxide and take away oxygen.

 c Look at the list of important features of a gas exchange system. Which parts of a fish's gas exchange system match the features in the list?

Investigating breathing

When you breathe in, your diaphragm flattens and your ribs are pulled upwards and outwards by the **intercostal muscles**. This increases your chest volume and lowers the pressure inside your lungs. Air is forced into your lungs by atmospheric pressure.

 d What is your breathing rate at rest? Give your answer in breaths per minute.

When you breathe out, the elasticity of your lungs pushes most of the air out of them. Even when you breathe out as much as possible some air always stays in your airways. This is your **residual air**. Changes in your breathing rate and lung capacity can be measured and recorded at rest and during exercise using a machine called a **spirometer**.

The spirometer absorbs the carbon dioxide from your breath enabling it to measure the efficiency of gas exchange. The air entering and leaving your lungs at rest is called your **tidal air**.

During exercise, you exchange more air by breathing deeper and faster. The maximum volume of air you can take in and breathe out in one large breath is called your **vital capacity**. The total lung volume is calculated by adding the vital capacity and residual air together. For most adults this is between 3 and 6 dm^3.

e Look at the spirometer trace. Use it to estimate the person's:
- tidal volume
- vital capacity.

f Use your answer to **d** to calculate how much air you exchange in 1 minute at rest. Assume your tidal volume is the same as the person above.

▲ A spirometer being used to measure the ventilation of a person's lungs

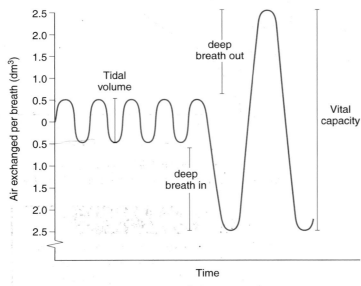

▲ A spirometer trace at rest followed by a deep breath in and out

Gasping for breath

Sometimes things go wrong with the gaseous exchange system. The cause of the problem may be genetic, or may be due to breathing in harmful substances or as a result of infection. Because the lungs have a large surface area and consist of millions of little sacs, it is important that harmful substances are prevented from entering and damaging them. The respiratory system is prone to diseases because the lungs are a 'dead end'.

The trachea and bronchi are lined with cells, called goblet cells, which secrete mucus, and ciliated cells that waft mucus away from the lungs. Mucus traps dust, spores, bacteria and viruses. The mucus passes into the throat and is swallowed. Stomach acid kills most micro-organisms.

Cystic fibrosis (CF) is a genetic disorder where the goblet cells make mucus that is unusually sticky. The cilia cannot remove the mucus which clogs the lungs and makes them less efficient and prone to infections.

Asbestosis is an industrial disease caused by inhaling fibres of asbestos. Asbestos was once used for insulation. Tiny particles, released when asbestos is cut, enter the lungs and cause breathing difficulties. Many years later this may develop into cancer.

Tar in tobacco smoke contains a number of carcinogens (cancer inducing chemicals) which cause cells to divide uncontrollably to form **lung cancer**. Smoking can also cause bronchitis (loss of cilia and over-secretion of mucus in the airways), emphysema (loss of alveoli) and **asthma**.

Cigarette smoking in men became popular in World War 1 (1914–1918). In the 1940s many women also took up smoking. Sales of cigarettes increased rapidly in the first half of the last century.

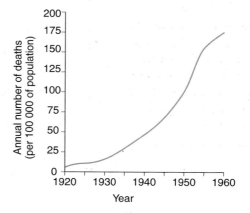

▲ Deaths from lung cancer in England and Wales

 g Look at the graph at the top of the page of deaths from lung cancer.
   (i) How does the graph suggest a link between smoking and cancer?
   (ii) How does the graph support the idea that the effects of smoking are long-term?

h Look at the graph on the right of deaths from lung cancer in men.
   (i) Jack smokes 15 cigarettes a day and Mike smokes 30 cigarettes a day.
      What are their increased risks of dying from lung cancer?
   (ii) Use the graph to suggest why some people think that smoking should be banned in public places.

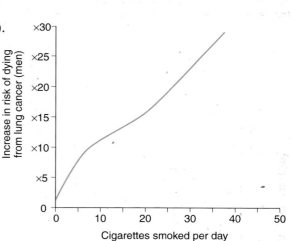

▲ The number of deaths from lung cancer in men varies with the number of cigarettes smoked per day

Keywords

asbestosis • asthma • cystic fibrosis • gill • intercostal muscle • lung cancer • residual air • spirometer • tidal air • vital capacity

Predicting asthma attacks

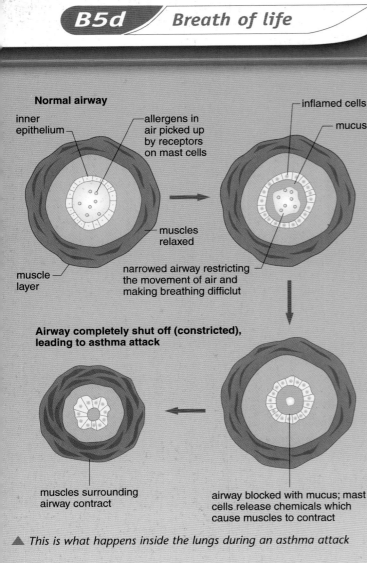

Normal airway

inner epithelium

allergens in air picked up by receptors on mast cells

muscles relaxed

muscle layer

narrowed airway restricting the movement of air and making breathing difficlut

inflamed cells

mucus

Airway completely shut off (constricted), leading to asthma attack

muscles surrounding airway contract

airway blocked with mucus; mast cells release chemicals which cause muscles to contract

▲ *This is what happens inside the lungs during an asthma attack*

Monnika has been given a pet rabbit. Every time she handles her rabbit, Monnika feels a tightness in her chest, begins to cough and has difficulty breathing. Monnika sees her doctor who says she has asthma and that Monnika's symptoms are caused by a narrowing of the tiny air passages in her lungs.

Asthma affects more than one in ten children. The cause is often an allergen, such as animal fur, that triggers an asthma attack. Most asthmatics know what their allergens are. Common allergens are feathers, pollen, dust mites, cigarette smoke, fungal spores, perfumes or pesticides.

The diagram shows what happens during Monnika's asthma attack. She uses a peak flow meter to check the condition of her lungs. The peak flow meter measures how efficiently her lungs empty. This measurement depends on the diameter of her airways. Low readings are an indication of narrow airways.

Monnika uses an inhaler to relax the muscles around her airways and takes medicines to inhibit allergic reactions.

Questions

Look at the diagram showing what happens during an asthma attack.

1 Mast cells detect the presence of particles in the airways. Suggest why some people's asthma is triggered by pollen and others' is triggered by cigarette smoke.

2 What is the function of the mucus secreted by the cells lining the airways?

3 Some drugs used to treat asthma block the receptors on the mast cells. How might this help to reduce the effects of asthma?

4 Suggest why the drugs that relax the muscles around the airways are called bronchodilators.

Use the graphs of peak flow values to help you to answer these questions.

5 How do the peak flow readings of an uncontrolled asthmatic differ from those of a normal subject?

6 What effect does medication have on a controlled asthmatic?

7 When do asthmatics usually have their lowest peak flow readings?

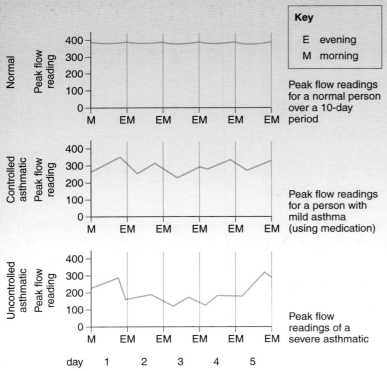

Key

E evening
M morning

Peak flow readings for a normal person over a 10-day period

Peak flow readings for a person with mild asthma (using medication)

Peak flow readings of a severe asthmatic

▲ *How peak flow values vary*

Dealing with waste

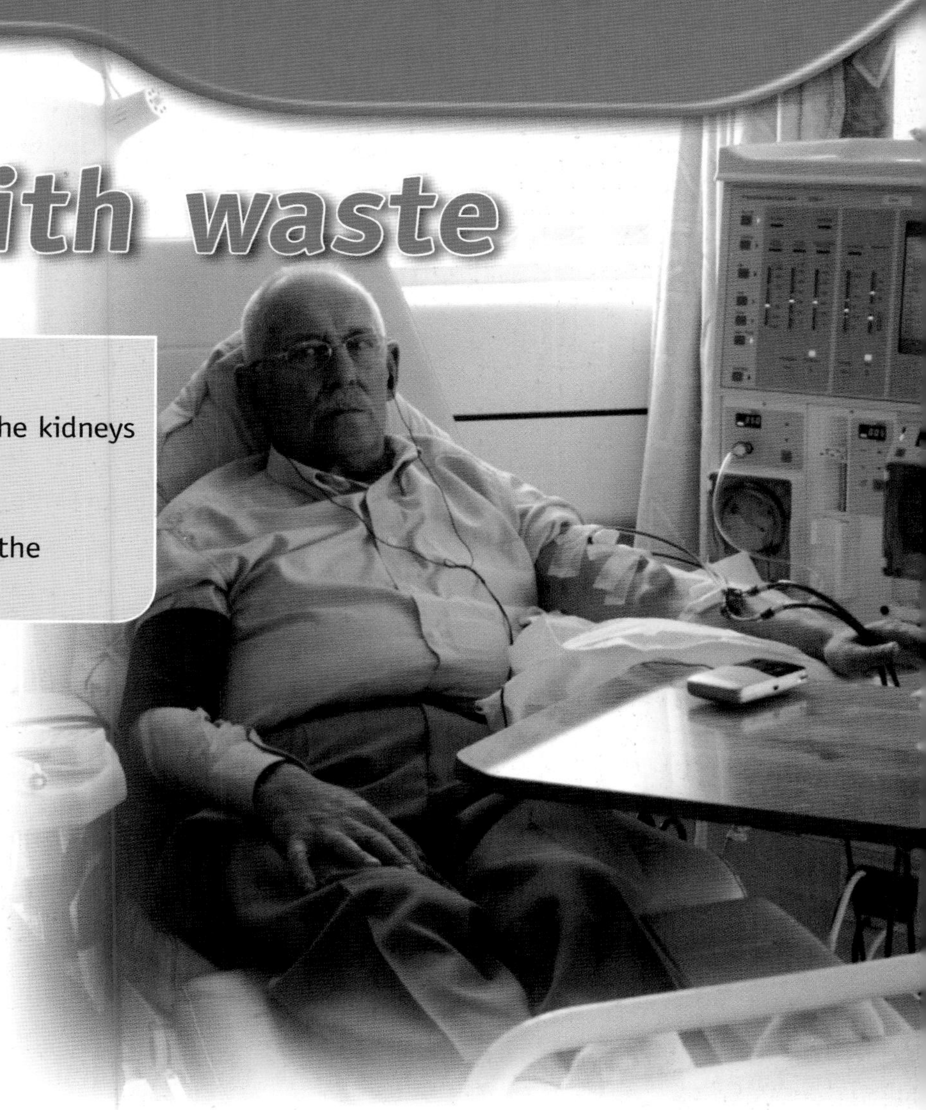

In this item you will find out

- about the structure and function of the kidneys
- how urine is formed in the kidneys
- how the lungs and the skin help in the removal of waste products

Your body consists of billions of active cells performing thousands of chemical reactions. These reactions produce toxic wastes that must be removed by a process called excretion. The organs that remove waste products are the **kidneys**, the **liver**, the lungs and the skin.

Dennis' kidneys have stopped working. Each week Dennis spends several hours on a kidney **dialysis** machine. The dialysis machine 'washes' Dennis' blood and removes the toxic waste. He will need to use this machine until a suitable kidney is available for transplant surgery.

▲ A kidney patient receiving dialysis to remove waste from his blood

a Explain why Dennis must have regular dialysis treatment.

Cellular activity makes three main waste products: urea, carbon dioxide and water. Urea is produced when the liver breaks down excess **amino acids**. Urea is removed by the kidneys.

Carbon dioxide and water are products of respiration. The carbon dioxide level in your blood is detected by nerve endings in your brain and major arteries. When the carbon dioxide level rises, a control centre in your brain increases your rate and depth of breathing. This increase continues until your blood's carbon dioxide level is returned to normal. When the CO_2 level in the lungs exceeds 5%, it must be removed.

Cellular activity involves chemical reactions that produce heat. Your skin helps to regulate your body temperature. When you are hot, the extra heat energy is used to evaporate sweat. Water molecules containing heat energy escape, cooling the skin.

b Why does cellular activity produce heat?

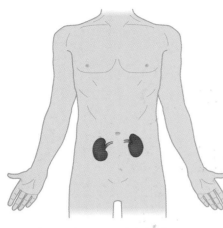

▲ *The location of the kidneys*

The kidney – a balancing act

Imagine an organ about the size of your fist. Inside this organ are about a million fine **tubules** which, unravelled, would stretch over 60 km. Each tubule is surrounded by an even longer tangle of capillaries that carry nearly 1000 litres of blood each day. The tubules and their blood capillaries are called nephrons and the fist-sized organs are your kidneys.

Your kidneys are located in your abdomen on either side of the aorta. Every day, your blood passes through the kidneys about 400 times. Urea, excess salt and water are removed from your blood, stored in the bladder, and excreted as **urine**.

c Suggest why blood needs to pass through your kidneys over 400 times each day.

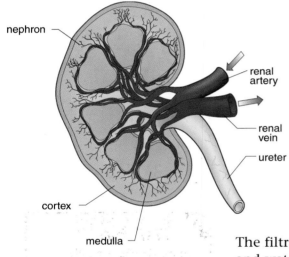

▲ *Inside a kidney*

How do the kidneys work?

The diagram opposite shows the three main areas of a nephron. Blood enters the nephron from the renal artery under pressure. It passes through a filter unit (labelled A) which consists of a knot of capillaries. This is called the **glomerulus** and it is where small molecules are filtered out of the blood under high pressure. This is called ultra-filtration.

A cup-shaped **capsule** collects the filtrate and passes it into the kidney tubule (labelled B). The first part of the tubule is folded and surrounded by more capillaries. Glucose is removed from the filtrate and put back into the blood by a process called **selective reabsorption**.

The filtrate continues through the kidney tubule to a region where salt and water are regulated. Here more capillaries reabsorb water and salts back into the blood. When the filtrate finally leaves the nephron at the point labelled C, all of the glucose and most of the salt and water has been removed leaving a solution containing urea. This is urine.

The table shows the percentage composition of plasma, filtrate and urine.

Substance	Blood plasma at A	Filtered fluid at B	Urine at C
Water	90	99	97
Proteins	9	0	0
Glucose	0.1	0.1	0
Urea	0.03	0.03	2.0

d Look at the table. Describe the difference between the percentage of protein in the plasma at A and in the filtrate at B. Suggest a reason for the difference.

e Suggest an explanation for the increase in the percentage of urea in the urine at C.

Amazing fact

Some foods alter the colour of your urine. Asparagus can turn it green, beetroot can turn it pink.

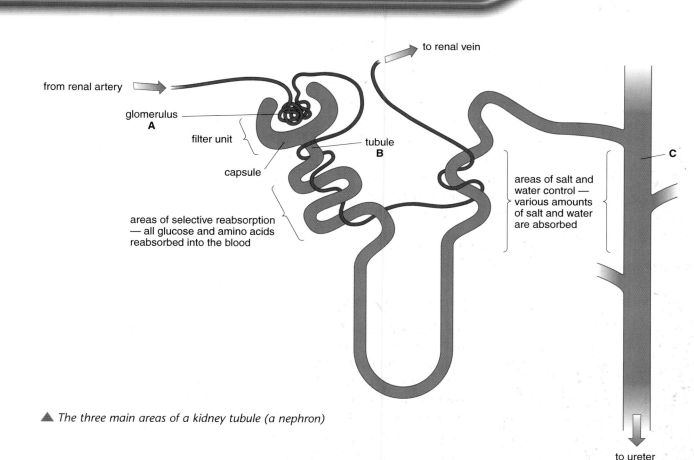

from renal artery

to renal vein

glomerulus
A

filter unit

tubule
B

capsule

C

areas of salt and
water control —
various amounts
of salt and water
are absorbed

areas of selective reabsorption
— all glucose and amino acids
reabsorbed into the blood

to ureter

▲ *The three main areas of a kidney tubule (a nephron)*

Controlling urine production

Your body controls its water content by balancing its intake and output.
The amount of water vapour you lose in breathing and sweating is
unavoidable. It varies with activity and temperature. If you lose too much
water, you feel thirsty and drink to replace the water that you have lost. If
you do not drink, your **pituitary gland** releases **ADH (anti-diuretic
hormone)** which makes your kidneys reabsorb more water.

Part of your nephron reabsorbs salt into the blood. This makes your
blood more concentrated and creates an osmotic gradient. ADH makes
your tubules more permeable to water so water passes into your blood by
osmosis. When your blood is too dilute, the production of ADH is switched
off. The tubules become less permeable to water so your urine output
increases. This type of control is an example of **negative feedback**.

f (i) **Which hormone controls urine output?**
(ii) **Where is this hormone made?**
(iii) **What sort of conditions stimulate the production of this hormone?**

Amazing fact

**Some drinks actually
dehydrate you!
Alcoholic drinks and
strong coffee both
block the production of
ADH. This means that
your kidneys produce
more urine.**

Keywords

ADH (antidiuretic
hormone) • amino acids •
capsule • dialysis
• glomerulus • kidney •
liver • negative feedback •
pituitary gland • selective
reabsorption • tubule •
urine

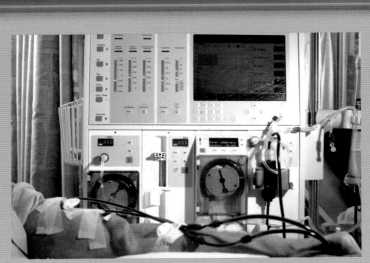

▲ *Dennis needs dialysis two or three times a week until a suitable kidney can be found*

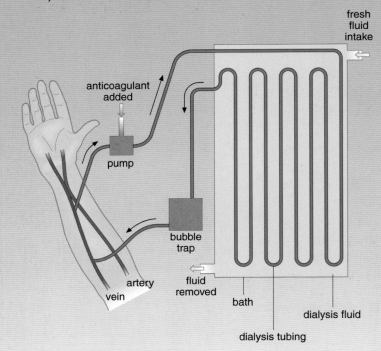

fresh fluid intake

anticoagulant added

pump

bubble trap

artery
vein

fluid removed

bath

dialysis fluid

dialysis tubing

▲ *Inside a kidney dialysis machine*

Dennis' story

Dennis had felt unwell for some time. He visited his doctor who did some routine blood tests. The results of the tests showed raised levels of potassium and urea. Eventually a diagnosis of kidney disease was made. Dennis was put on a low potassium diet avoiding citrus fruits, bananas, coffee and chocolate, amongst other things.

Despite this, Dennis still felt unwell and his kidneys continued to fail. Dennis was told that he would need dialysis to do the work of his kidneys. He was given the Giovanetti diet, limiting his protein intake to about 40 g per day to reduce urea production. His fluid intake was reduced to about 1 pint of liquid per day. Dennis now has dialysis three times a week. Each treatment lasts about 6 hours.

The dialysis machine takes blood from a vein in Dennis' arm. An anticoagulant is added to stop Dennis' blood clotting. His blood passes through dialysis tubing, which is selectively permeable and made of cellophane. The tubing is bathed in dialysis fluid with a similar composition to normal plasma. Dennis' blood is 'washed' by this fluid which allows waste products and excess water to diffuse out of his blood and into the dialysis machine.

The sodium and glucose balance of Dennis' blood is maintained by controlling the concentration of these substances in the dialysing fluid. Before the dialysed blood is returned to Dennis' arm it is passed through a filter to remove any air bubbles or tiny clots that could block his circulation.

Eventually Dennis hopes to have a kidney transplant. He will then be freed from the inconvenience of dialysis and will be able to lead a normal life once again.

Questions

1 Explain why the protein content of Dennis' diet is limited to 40 g per day.

2 Why is it necessary to add an anticoagulant to the patient's blood?

3 Suggest how dialysis removes urea and potassium from the blood while keeping the concentrations of glucose and other salts in the blood at normal levels.

4 During dialysis, Dennis is able to drink extra fluids. Suggest why.

5 Suggest possible reasons why Dennis is still waiting for a kidney transplant.

Making babies

▲ *Carrying on the human race*

In this item you will find out

- about the main stages of the menstrual cycle and how it is controlled by hormones

- what options are open to infertile people who want children

- how the development of the fetus can be monitored during pregnancy

Like other animals, humans need to reproduce to carry on our species. Every day thousands of babies are born all over the world. Most people get pregnant naturally and go on to have healthy babies, but sometimes there are problems.

Deciding when to have children, and how many, is one of life's most important decisions. Many couples decide to put off having children until they have travelled, completed training for their careers or become financially secure.

Avoiding unplanned pregnancies is one way of delaying becoming a parent, but then the likelihood of complications increases with age. Above the age of 30, a woman's fertility decreases and her chances of having a child with **Down's syndrome** increases. A recently developed test enables a woman to measure her fertility by means of a simple urine test. This test measures her hormone levels and can indicate when her fertility is starting to decline.

Even healthy young couples can suffer from **infertility**. Advances in medical science now mean that this can be treated in several ways. But not everyone thinks that such treatments are a good idea.

Babies can be screened in the womb to check whether they have conditions such as Down's syndrome. If there is a problem then the woman can choose to terminate the pregnancy. This raises ethical issues.

Amazing fact

Fertility drugs can extend a woman's reproductive life. Using hormones it is possible for a woman in her sixties to become a mother.

 a A woman has a reproductive life of 40 years, releasing an egg every 28 days. She has two children from two pregnancies. How many eggs will she have released in total by the end of her reproductive life?

 b Suggest why it is particularly important to screen women who have become pregnant as a result of fertility treatment.

The menstrual cycle

Most women menstruate (have periods) once every 28 days. The **menstrual cycle** is shown in the diagram.

The uterus lining thickens so it is ready to receive an egg. Around day 14, an egg is released by the ovary. If the egg is not fertilised, the uterus lining breaks down and eventually passes out of the vagina as a 'period' two weeks later. The cycle then begins all over again. This cycle is controlled by four hormones: oestrogen, progesterone, **follicle stimulating hormone (FSH)** and **luteinising hormone (LH)**.

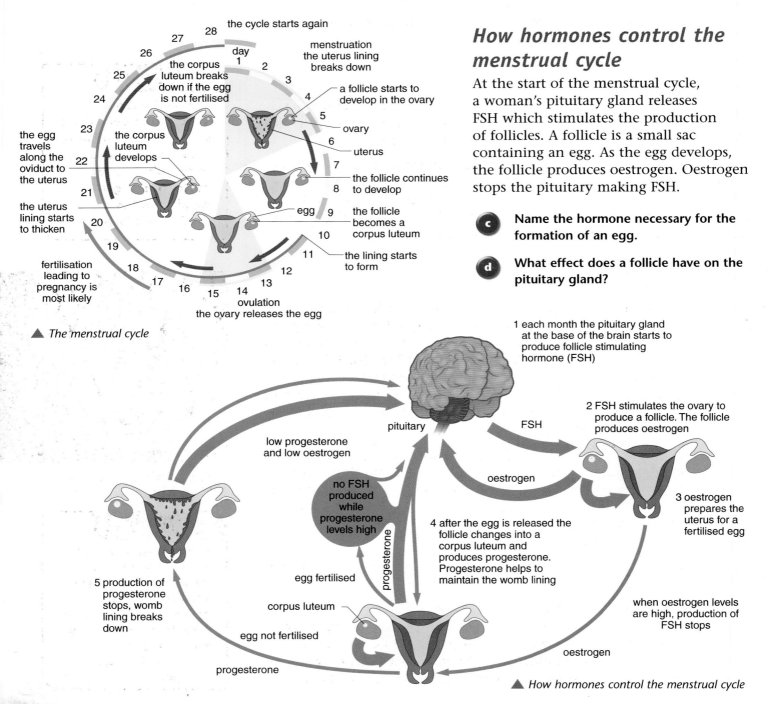

the cycle starts again

menstruation the uterus lining breaks down

the corpus luteum breaks down if the egg is not fertilised

a follicle starts to develop in the ovary

ovary

the corpus luteum develops

uterus

the egg travels along the oviduct to the uterus

the follicle continues to develop

the uterus lining starts to thicken

egg

the follicle becomes a corpus luteum

fertilisation leading to pregnancy is most likely

the lining starts to form

ovulation the ovary releases the egg

▲ The menstrual cycle

How hormones control the menstrual cycle

At the start of the menstrual cycle, a woman's pituitary gland releases FSH which stimulates the production of follicles. A follicle is a small sac containing an egg. As the egg develops, the follicle produces oestrogen. Oestrogen stops the pituitary making FSH.

c Name the hormone necessary for the formation of an egg.

d What effect does a follicle have on the pituitary gland?

1 each month the pituitary gland at the base of the brain starts to produce follicle stimulating hormone (FSH)

pituitary

FSH

2 FSH stimulates the ovary to produce a follicle. The follicle produces oestrogen

low progesterone and low oestrogen

oestrogen

no FSH produced while progesterone levels high

3 oestrogen prepares the uterus for a fertilised egg

progesterone

4 after the egg is released the follicle changes into a corpus luteum and produces progesterone. Progesterone helps to maintain the womb lining

egg fertilised

5 production of progesterone stops, womb lining breaks down

corpus luteum

when oestrogen levels are high, production of FSH stops

egg not fertilised

oestrogen

progesterone

▲ How hormones control the menstrual cycle

After she ovulates (releases an egg) the woman's pituitary gland produces LH. This makes her follicle develop into a gland called a corpus luteum. This gland makes progesterone. If the egg is not fertilised, the gland shrinks and stops making progesterone. This triggers menstruation.

During pregnancy, progesterone continues to be made, maintaining the uterus lining and preventing the production of FSH. This is shown in the diagram.

 e Which structure produces progesterone?

f (i) Describe what happens to the levels of oestrogen and progesterone at the end of the menstrual cycle.
(ii) What effect do these levels of oestrogen and progesterone have on the pituitary gland?

Screening babies

Fetal screening is used to monitor pregnancies. Pregnant women have ultrasound scans that produce moving images of the fetus. Scans can detect a baby's sex, multiple pregnancies or developmental defects.

Cells in the liquid surrounding the fetus are sampled using a hypodermic needle to remove some of the amniotic fluid. The cells are checked for chromosome abnormalities, such as Down's syndrome. This is called **amniocentesis**. Rapid advances are being made in the early diagnosis of genetic disorders that affect the health or development of the fetus.

Treating infertility

The world's first test-tube baby, Louise Brown, was born in 1978. Her mother had blocked oviducts so doctors collected eggs from her ovary. On 10 November 1977, an egg was fertilised in a glass dish and grown into an embryo. This technique is called **IVF (in-vitro fertilisation)**. When her mother's hormones were at the correct levels, the embryo was placed in her uterus and left to develop to full term. IVF is now a common treatment for women with blocked oviducts. There are thousands of people who have been conceived by this method.

Women who fail to ovulate may be given FSH. Early attempts at this treatment often stimulated the release of several eggs producing multiple births. If FSH treatment fails, eggs from a donor can be fertilised with the man's sperm using IVF before they are put in the infertile woman's uterus. Women whose ovaries have been damaged by accident or disease are sometimes given an ovary transplant.

A **surrogate** mother can be implanted with a couple's embryo when the biological mother is unable to have a normal pregnancy.

Inability to produce sufficient sperm is a cause of infertility in men. A man's sperm count can be increased by 'spinning' semen in a centrifuge to concentrate it. The concentrated sperm is placed inside the woman's uterus. This is a form of **artificial insemination** called IUI (intra-uterine insemination). If the man makes no sperm, donated sperm can be used.

▲ *Louise Brown, the world's first test-tube baby*

Keywords

amniocentesis • artificial insemination • Down's syndrome • fetal screening • follicle stimulating hormone (FSH) • infertility • IVF (in-vitro fertilisation) • luteinising hormone (LH) • menstrual cycle • surrogate

Problems conceiving

▲ *There are several ways to help couples overcome infertility problems*

Angie and Joe have wanted to start a family for two years but without success. They have been referred to a consultant who has carried out some tests. Earlier, the consultant explained that about half of the causes of infertility are a result of problems with both the man and the woman. A quarter of the cases involve just the man.

Joe has given three samples of his semen for testing and is waiting to hear the results.

The consultant tells Joe and Angie that all three of Joe's tests show some abnormalities. His semen has a low concentration of sperm cells and many of his sperms move slower than is usual.

The consultant tells them that there are several treatment options. One option would be IUI where some of Joe's semen would be placed high in the uterus when Angie has just ovulated. This would increase the chance of successful fertilisation.

Another option is to concentrate some of Joe's semen to improve the sperm count. A hormone would be used to stimulate Angie's ovaries before she is artificially inseminated with Joe's semen. This is AIH/S (artificial insemination by husband with stimulation).

A third option would be to use IVF or ICSI. ICSI is short for Intracytoplasmic Sperm Injection. One of Joe's active sperms would be injected directly into one of Angie's eggs. This treatment is expensive, but has the best success rate with a 70% cumulative chance of a live birth after four treatment cycles.

Finally, Angie could be fertilised using sperm from a donor.

Angie is 35. They decide to choose the ICSI option. They do this because Angie's fertility is likely to drop in a few years and there is a greater risk of complications and genetic abnormalities when a woman nears the end of her reproductive life.

Questions

1 Suggest how each of the following problems lowers Joe's chances of becoming a father:
   (a) a low sperm count
   (b) poor sperm mobility.

2 Explain why inserting sperm high into the uterus after ovulation increases the chance of successful fertilisation.

3 Both IVF and ICSI involve the sperm and egg joining outside the woman's body. How could this be used to check that the egg is successfully fertilised?

4 ICSI is an expensive procedure. Should this treatment be available to all couples with infertility problems? Justify your answer.

5 If ICSI treatment is to be limited, who should receive it and who should decide?

6 Suggest why some people object to any form of fertility treatment.

New for old

In this item you will find out

- about the problems of using mechanical replacements for body parts

- about the problems of finding suitable donor organs and preventing rejection of donated organs

- about some of the ethical issues involved in transplant surgery

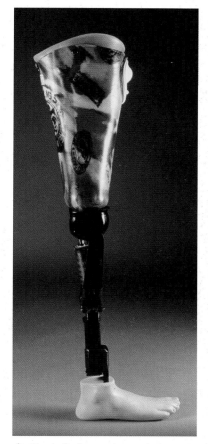

▲ Bionic limbs are becoming increasingly sophisticated

Are you a fan of classic horror stories? The horror story *Frankenstein* is about a scientist who builds a monster from body parts. In the nineteenth century, there was an industry in obtaining bodies. The 'body-snatchers' Burke and Hare dug up fresh corpses and sold them to anatomists for dissection. There was a thirst for finding out about the disease process and how damaged organs might be restored. Some futuristic stories, like *The Bionic Man*, use ideas of part human, part machine as their theme.

As people live longer, their body parts are more likely to wear out or go wrong, and more people will seek treatments. Body part replacements currently range from false teeth to artificial limbs, and from skin grafts to heart transplants.

The scope for replacement surgery is growing. For example, face transplants may replace plastic surgery for some people scarred by burns or disease. This raises some important issues. The families of most organ donors usually feel happy about someone benefiting from the death of their loved ones. The people who receive the donated organs are usually extremely grateful. However, people who have undergone plastic surgery sometimes feel uncomfortable with themselves and find it difficult to adjust. A person receiving someone else's face may find the psychological recovery difficult. Their friends and family, and those of the donor, may also find this difficult to deal with.

 a Suggest one advantage and one disadvantage of replacing damaged body parts with artificial parts rather than with organs from a donor.

Amazing fact

In the nineteenth century human teeth were transplanted on to the head of a cockerel. The blood supply became connected and, for a while, the teeth survived until they were rejected by the cockerel's immune system.

I USE MACLUCKS TOOTHPASTE!

▲ You can still run with an artificial leg!

▲ This patient, undergoing major heart surgery, is attached to a heart-lung machine

Mechanical and electronic replacements

Body parts can be replaced in different ways. Mechanical devices, called implants, are made from a variety of substances that are carefully chosen for their properties. Implants must be made of strong, non-toxic materials that do not wear, corrode or cause allergic reactions. New lightweight plastics, ceramics and alloys have the necessary properties to make implants last much longer.

Most implants are engineered devices such as replacement joints, limbs, artificial heart valves and replacement lenses. Others are electronic devices such as pacemakers or hearing aids.

One of the problems of producing suitable implants is making them small enough to fit inside the body. Nanotechnology (micro-engineering) and advances in micro-electronics are rapidly solving the problems of miniaturisation. Another problem is providing enough reliable power to drive these devices. This is being overcome by using high performance batteries that can be recharged from outside the body. These are now commonly used in heart pacemakers.

Using machines

Sometimes mechanical and electronic implants cannot be used. When an organ suddenly begins to fail or if a person is seriously ill or injured, it may be possible to use a machine to do the work of the damaged organ. These machines are used outside the body until the person recovers or a suitable organ is found for transplant.

One of the earliest machines was the **iron lung**. Polio can weaken or paralyse the lung muscles making breathing difficult. An iron lung is a pressure chamber that fits around the chest. Air is pumped in and out of the chamber causing the chest to compress and expand, and this ventilates the lungs.

Other machines that are used are **heart and lung machines** and kidney dialysis machines.

b Explain how an 'iron lung' works.

Organ donation

Mechanical replacements can be used for some body parts, such as hip joints, eye lenses and heart valves, but some body parts need to be replaced with organs donated from other people. These donor parts include kidneys, lungs, corneas and livers.

Organs for transplantation must be healthy and undamaged, the right size and age and the correct tissue type. Donated organs must be implanted as quickly as possible so they do not deteriorate. Because of these requirements, and because of religious or cultural reasons, there is a shortage of suitable organs.

Some people carry donor cards while others register this wish on a computerised National Register of Donors. The Anthony Nolan Trust keeps a list of people who are willing to donate specific tissues such as bone marrow.

▲ Many people carry a donor card stating that their organs may be donated to someone else after they die

c Because there is a shortage of suitable donated organs, what factors do you think should be used to decide which patients will receive them?

d Hospital waiting lists for transplants are generally much longer than for implants. Suggest a reason for this.

Problems with transplants

If the transplanted organ carries the wrong antigens, it triggers an attack by the recipient's white blood cells. This is known as the immune response. The white cells, called T cells, begin the rejection process. An early sign of **rejection** is the destruction of the blood vessels in the transplanted organ. The death of cells and the formation of scar tissue follow this. Eventually the transplanted organ dies.

In the early 1980s, a drug, called cyclosporine, was isolated from a fungus. Cyclosporine is an **immuno-suppressive drug** which stops T cells attacking and rejecting the transplant. Since then the success rate of some transplants has risen to over 90%. This is the standard treatment to prevent organ rejection, but it also reduces the body's ability to fight infection. To prevent rejection, the recipient has to take immuno-suppressants for life, but the dose can usually be reduced over time.

In the late 1980s, drugs that attack individual T cells were developed. One of these drugs is used to treat donated bone marrow before it is put into the recipient. This prevents rejection of the bone marrow even if the donor and recipient are not closely matched. Using these drugs avoids the need to take immuno-suppressive drugs altogether.

e How might the destruction of blood vessels lead to the rejection of a transplanted organ?

f Suggest why a transplant between identical twins has a better chance of success than a transplant between brother and sister.

Examiner's tip

Close relatives are more likely to be suitable organ donors because they share many of the same genes and so have more similar tissue types.

Amazing fact

In 2006, when this book was written, 6681 people were waiting for organ transplants in the UK.

Keywords

heart and lung machine • immuno-suppressive drug • iron lung • rejection

Multiple transplants

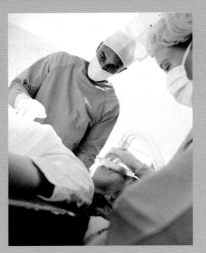

▲ *Maxine is being prepared for her heart-lung transplant*

Maxine is in her mid twenties. She has cystic fibrosis, a genetic disorder that has damaged her lungs. For most of her life she has taken regular medication and undergone physiotherapy but her condition has recently worsened. She is being prepared for a transplant operation.

Maxine's doctors will replace her heart and lungs with organs taken from a donor who has suffered fatal head injuries in a car accident. Maxine's doctors have matched her tissue type with that of the donor, who was a fit and healthy 27-year-old of similar size to Maxine. The donor's family have also given consent for more of his organs, including his kidneys and liver to be used for other transplant operations.

Maxine has also agreed to be a living donor. Although cystic fibrosis has affected her lungs, her healthy heart is suitable for transplanting into another patient with a defective heart. Although some people are reluctant to be organ donors, Maxine and her family have thought a great deal about this and are convinced of the potential benefits of this type of surgical treatment.

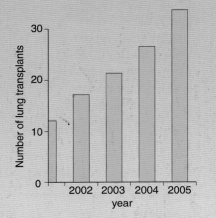

▲ *Bar chart showing the number of lung transplants performed in a large US hospital*

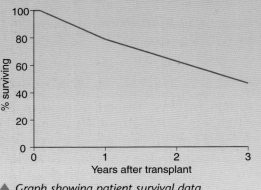

▲ *Graph showing patient survival data*

The bar chart shows the number of lung transplants performed in a large US hospital and the line graph shows patient survival data.

Questions

1 Suggest three factors that must be taken into account when selecting organs for transplantation.

2 Explain why Maxine's heart will be used for a second transplant although her lungs will not.

3 Some people are reluctant to become an organ donor. Suggest two reasons why.

4 Look at the bar chart.
   (a) How has the number of lung transplant operations changed?
   (b) How many lung transplants did the hospital perform up to 2005?

5 Look at the graph. What percentage of patients with lung transplants survived for:
   (a) 1 year
   (b) 2 years
   (c) 3 years?

Patterns of growth

In this item you will find out

- that the height of humans can be influenced by genes and hormones

- that environmental factors such as diet, exercise and disease can affect growth

- how patterns of growth can be monitored to highlight problems of development

▲ *Elephants have limited growth*

Have you ever wondered what limits the size of an organism? Why don't we grow to the size of a blue whale? Obviously we do not have the genetic plan to grow to this extreme of size. Have you noticed that there are different patterns of growth?

As an elephant grows, its proportions change. Different parts of the elephant's body grow at different rates. This is called limited growth. Crocodiles have unlimited growth. Crocodiles continue to grow slowly throughout life and their proportions do not change so noticeably. Insects grow in bursts, they have to shed their exoskeletons at each stage. In all of these organisms, growth is co-ordinated by chemicals.

▲ *Crocodiles have unlimited growth*

Most animals begin life as a single cell. New cells for growth are produced by a process called mitosis. This creates identical cells. Before a cell divides by mitosis, its genetic information is copied. Each daughter cell will have the same genetic information as the cell that produced it, but as the organism grows, the cells become specialised. This is because the growth and development of the cells is controlled by hormones and is affected by the availability of nutrients.

An animal that is fit and well fed will grow to its full potential. Animals with an imbalance of hormones may become dwarfs or giants. Animals with a poor diet may become thin or obese. Exercise also plays an important part in ensuring that an organism grows and develops normally.

 a What type of growth do you think humans show? Give reasons for your answer.

 b Why must insects grow in bursts?

Amazing fact

In the first 17 years of your life your mass increases by around 50 million million times. In the 9 months before you are born your mass increases over a million million times!

▲ *Robert Pershing Wadlow – the world's tallest man*

Measuring growth

Growth is usually measured as an increase in mass or height. Mass is quick and easy to measure, although it fluctuates. We weigh more if we have just eaten or have a full bladder. Similarly, our weight decreases when we are dehydrated. Before a fight or a race, boxers and jockeys sometimes 'sweat off' their weight in a sauna in order to meet their weight at the weigh-in. In babies, where growth is rapid, these variations in weight are relatively minor.

Measuring the mass, length and head size of a baby and comparing these measurements with **average growth charts** can help us to spot potential problems in growth. These measurements are made regularly during the baby's first few months. These average growth charts are only a guide as to how the baby should be growing. Some babies may grow at slower or faster rates.

Measuring height is easy if the person is able to stand. Babies are laid flat with their legs extended to measure their length. Because parts of the human body grow at different rates, measuring height or length alone is not enough to check the development of a baby. To do this you must compare the proportions of different parts. In some conditions the head can grow too quickly or the limbs too slowly.

c It is usual to make regular measurements of a baby's length, head size and mass. Suggest why these three measurements are necessary.

Interpreting human growth

The graph on the left shows typical height curves for girls aged 2 to 18 years. The normal range for height is coloured green. If a girl's height is outside this area it may mean she is growing abnormally and she may have a hormonal or developmental problem.

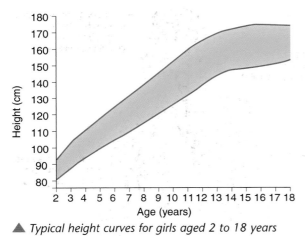

▲ *Typical height curves for girls aged 2 to 18 years*

d Niamh is 8 years old and 145 cm tall. What does this suggest about her rate of growth?

The graph below shows the typical growth in height of girls compared to boys from 5 to 19 years.

e Look carefully at the growth patterns of girls and boys. At which of the four ages are:
- girls taller than boys
- boys taller than girls?

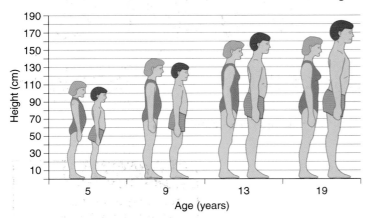

▲ *Combined graph of boys' and girls' heights*

Giants and dwarfs

Extremes of size grab the attention. The tallest man in the world that we know about was Robert Pershing Wadlow (1918–1940). He grew to a height of 272 cm.

The world's shortest adult that we know about is Gul Mohammed of India. He was measured in 1990 and found to be 57 cm in height.

These extremes in height are caused either by genes or by a hormone imbalance. The pituitary gland produces **growth hormone** which, with the aid of the liver, increases the growth of both muscle and cartilage at the ends of long bones, such as limb bones. **Giantism** is usually caused by an overactive pituitary gland. **Dwarfism** can be caused either by an underactive pituitary or by the lack of a hormone called thyroxine. Thyroxine is made by the thyroid gland located in your neck.

▲ *Gul Mohammed – the world's shortest adult*

Other hormones affecting growth are sex hormones (oestrogen and testosterone), insulin made by the pancreas and androgens made by the adrenal glands. Stunted growth is now treated using genetically engineered growth hormone or anabolic steroids (artificial substances that stimulate muscle development).

The table shows data on the growth of a normal boy compared to a boy with growth hormone deficiency.

 f Suggest how the data in the table show the importance of comparing the height of a person against an average growth chart.

 g How soon can we see that one of these boys might need treatment with growth hormone?

Age in years	Height (cm)	Height with growth hormone deficiency (cm)
Birth	52	52
2	86	76
4	102	85
6	115	91
8	126	100
10	137	106.5
12	147	110
14	161	114
16	173	118
18	175	128
20	175	130

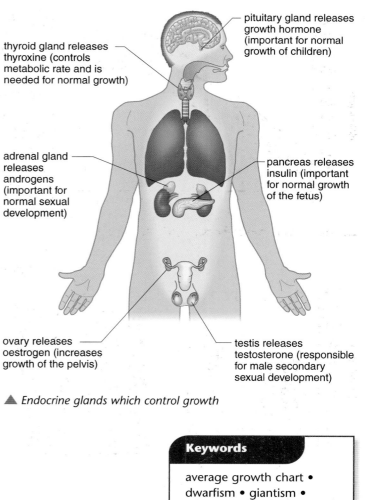

thyroid gland releases thyroxine (controls metabolic rate and is needed for normal growth)

pituitary gland releases growth hormone (important for normal growth of children)

adrenal gland releases androgens (important for normal sexual development)

pancreas releases insulin (important for normal growth of the fetus)

ovary releases oestrogen (increases growth of the pelvis)

testis releases testosterone (responsible for male secondary sexual development)

▲ *Endocrine glands which control growth*

Keywords

average growth chart •
dwarfism • giantism •
growth hormone

Healthy lives

▲ With people living longer, how will they cope in the future?

We all have only one life! Your average life expectancy depends on where you live. In more economically developed countries, people are living for longer. Increased wealth allows them to have better housing, better access to modern medical treatment and a healthier diet and lifestyle. Water and food supplies are safe and there are fewer industrial diseases caused by poor working conditions. Improved education and health awareness means people can reduce the risk of developing life-threatening conditions. A fit and healthy working population allows a society to flourish.

Because of this, one problem in more economically developed countries is an increasingly ageing population. A population with a large proportion of elderly people must spend more money on pensions, hospitals and social care.

Sandra is a pensioner who gets a state pension. She is 88 years old and still lives in her own house. The government has pledged to increase the amount of state pension that pensioners receive, which means that people in work will have to pay higher taxes. An ageing population means that more and more people receive state pensions so taxpayers will have to carry the burden.

Sandra has a daughter who lives several hours away, so she does not get to see her as often as she would like. A home help visits several times a week. This is also paid for by the government. Sandra is worried that she might have to go into an old people's home as she is finding it more and more difficult to get up and down the stairs to her bedroom and bathroom. She may have to sell her house to cover the cost of a home and she doesn't want to do that because she wants to leave it to her daughter.

Questions

1 What are the factors that contribute to increased life expectancy in more economically developed countries?

2 Suggest three things that an individual can do to improve their life expectancy.

3 Why should we be concerned about people living longer?

4 How does an ageing population affect taxpayers?

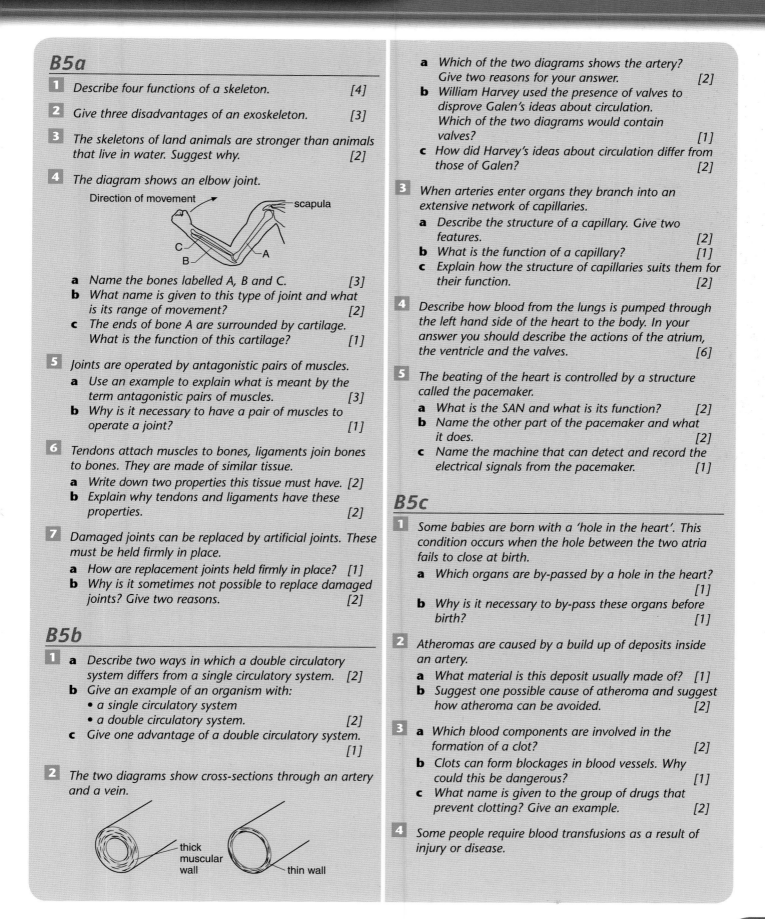

B5a

1 Describe four functions of a skeleton. [4]

2 Give three disadvantages of an exoskeleton. [3]

3 The skeletons of land animals are stronger than animals that live in water. Suggest why. [2]

4 The diagram shows an elbow joint.

Direction of movement
scapula
C
B
A

a Name the bones labelled A, B and C. [3]

b What name is given to this type of joint and what is its range of movement? [2]

c The ends of bone A are surrounded by cartilage. What is the function of this cartilage? [1]

5 Joints are operated by antagonistic pairs of muscles.

a Use an example to explain what is meant by the term antagonistic pairs of muscles. [3]

b Why is it necessary to have a pair of muscles to operate a joint? [1]

6 Tendons attach muscles to bones, ligaments join bones to bones. They are made of similar tissue.

a Write down two properties this tissue must have. [2]

b Explain why tendons and ligaments have these properties. [2]

7 Damaged joints can be replaced by artificial joints. These must be held firmly in place.

a How are replacement joints held firmly in place? [1]

b Why is it sometimes not possible to replace damaged joints? Give two reasons. [2]

B5b

1 **a** Describe two ways in which a double circulatory system differs from a single circulatory system. [2]

b Give an example of an organism with:
• a single circulatory system
• a double circulatory system. [2]

c Give one advantage of a double circulatory system. [1]

2 The two diagrams show cross-sections through an artery and a vein.

thick muscular wall
thin wall

a Which of the two diagrams shows the artery? Give two reasons for your answer. [2]

b William Harvey used the presence of valves to disprove Galen's ideas about circulation. Which of the two diagrams would contain valves? [1]

c How did Harvey's ideas about circulation differ from those of Galen? [2]

3 When arteries enter organs they branch into an extensive network of capillaries.

a Describe the structure of a capillary. Give two features. [2]

b What is the function of a capillary? [1]

c Explain how the structure of capillaries suits them for their function. [2]

4 Describe how blood from the lungs is pumped through the left hand side of the heart to the body. In your answer you should describe the actions of the atrium, the ventricle and the valves. [6]

5 The beating of the heart is controlled by a structure called the pacemaker.

a What is the SAN and what is its function? [2]

b Name the other part of the pacemaker and what it does. [2]

c Name the machine that can detect and record the electrical signals from the pacemaker. [1]

B5c

1 Some babies are born with a 'hole in the heart'. This condition occurs when the hole between the two atria fails to close at birth.

a Which organs are by-passed by a hole in the heart? [1]

b Why is it necessary to by-pass these organs before birth? [1]

2 Atheromas are caused by a build up of deposits inside an artery.

a What material is this deposit usually made of? [1]

b Suggest one possible cause of atheroma and suggest how atheroma can be avoided. [2]

3 **a** Which blood components are involved in the formation of a clot? [2]

b Clots can form blockages in blood vessels. Why could this be dangerous? [1]

c What name is given to the group of drugs that prevent clotting? Give an example. [2]

4 Some people require blood transfusions as a result of injury or disease.

a What precautions should be taken when selecting blood for a transfusion? [2]

b Explain why a person with group A blood can donate cells to a person with AB blood but not the reverse. [3]

B5d

1 Look at the diagram of the gills of a fish.

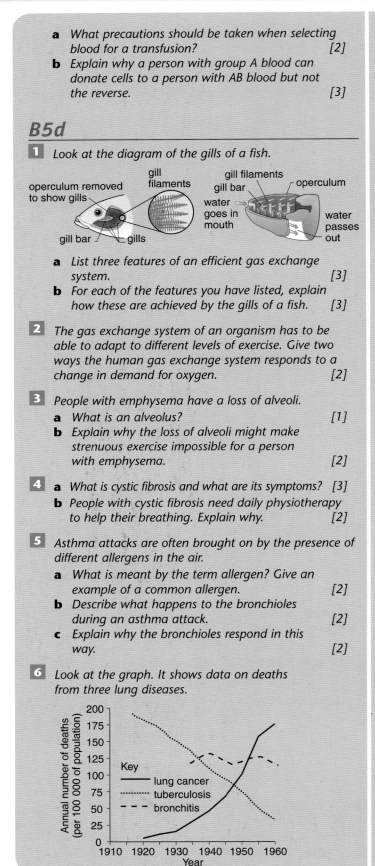

operculum removed to show gills

gill filaments

gill filaments

gill bar

operculum

water goes in mouth

water passes out

gill bar / gills

a List three features of an efficient gas exchange system. [3]

b For each of the features you have listed, explain how these are achieved by the gills of a fish. [3]

2 The gas exchange system of an organism has to be able to adapt to different levels of exercise. Give two ways the human gas exchange system responds to a change in demand for oxygen. [2]

3 People with emphysema have a loss of alveoli.

a What is an alveolus? [1]

b Explain why the loss of alveoli might make strenuous exercise impossible for a person with emphysema. [2]

4 **a** What is cystic fibrosis and what are its symptoms? [3]

b People with cystic fibrosis need daily physiotherapy to help their breathing. Explain why. [2]

5 Asthma attacks are often brought on by the presence of different allergens in the air.

a What is meant by the term allergen? Give an example of a common allergen. [2]

b Describe what happens to the bronchioles during an asthma attack. [2]

c Explain why the bronchioles respond in this way. [2]

6 Look at the graph. It shows data on deaths from three lung diseases.

Annual number of deaths (per 100 000 of population)

200
175
150
125
100
75
50
25
0

1910 1920 1930 1940 1950 1960
Year

Key
— lung cancer
····· tuberculosis
--- bronchitis

a Describe the patterns shown by the data on these three lung diseases. [3]

b Antibiotics were first produced around 1940. Tuberculosis of the lungs is caused by a bacterium. Bronchitis is an inflammation of the airways caused by dust and smoke. Use this information to suggest why the patterns of data for these two diseases are different. [2]

c In which ten year period was the rate of increase in deaths from lung cancer the greatest? [1]

B5e

1 **a** What is meant by the term excretion? [2]

b Name three substances that are excreted by cells. [3]

c Explain why excretion is necessary. [2]

2 **a** In which organ of the body is urea made? [1]

b Describe how urea is produced by this organ. [3]

c Which organs are responsible for removing urea from the blood? [1]

3 The diagram shows a part of an excretory organ which has been highly magnified.

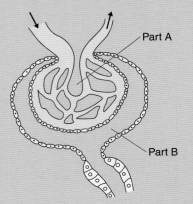

Part A

Part B

a In which excretory organ would you expect to find structures like this? [1]

b Part A consists of a bundle of capillaries. Name parts A and B. [2]

c Liquid leaks out of part A and is collected by part B. What name is given to this process? [1]

4 Kidney patients have to use a dialysis machine.

a What is meant by the term dialysis? [2]

b How will the composition of the patient's blood change after it has passed through the machine? [2]

c The dialysis fluid entering a machine has a similar concentration to blood plasma. Explain why. [1]

5 On a hot day, or after a long period of vigorous exercise, the production of urine decreases. Explain why. Your answer should include how body conditions are regulated including the role of ADH. [5]

B5f

1 Arrange the following 4 stages of the menstrual cycle in their correct order:

build up of uterus lining   development of a follicle
ovulation   period   [3]

2 Match each of these hormones to their correct function:

Hormone	Function
1 FSH	a Stimulates the development of a corpus luteum
2 LH	b Stimulates the development of an egg
3 Oestrogen	c Maintains the uterus lining during pregnancy
4 Progesterone	d Helps to repair the uterus lining after the period [3]

3 Describe what is meant by each of the following treatments for infertility:
- artificial insemination
- IVF
- FSH therapy. [3]

4 Look at the diagram. It shows a method of testing a fetus for conditions such as Downs Syndrome.

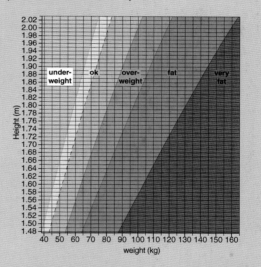

umbilical cord

a What name is given to this type of fetal screening? [1]

b Explain how this method of screening works. [3]

5 There are many different treatments available for couples with infertility problems. One method is IVF. Discuss the arguments for and against this method of treating infertility. [5]

B5g

1 List four factors that must be considered when choosing an organ for transplantation. [4]

2 Describe three problems of using mechanical replacements for damaged organs. [3]

3 One of the problems of using organ transplants is that of rejection.

a What is meant by the term rejection? [1]

b Explain why certain organs are rejected when they are transplanted into a patient. [3]

4 Some people think that there should be a national register of organ donors kept on a central computer.

a Write down three advantages of keeping a national register of donors. [3]

b Some people object to the idea of organ donation. Suggest two reasons why. [2]

5 People with damaged or diseased kidneys can be kept alive using a kidney dialysis machine. They could also be given a kidney transplant. Give an advantage and a disadvantage of each of these two types of treatment. [4]

B5h

1 Look at the graph. It shows human average height to weight ratios. Nutritionists use graphs like this to advise people how to eat healthily.

a Alastair is 175 cm tall, he weighs 50 kg. What does the graph tell you about Alastair's height to weight ratio? [1]

b What advice do you think the doctor should give to Alastair? Explain why. [2]

2 The growth of an organism involves the production of more cells. Name the type of cell division that produces an increase in the number of body cells. [1]

3 People in more economically developed countries are living longer on average.

a State three reasons for this increase in average life expectancy. [3]

b Write down two problems facing a country where the average age of the population is increasing. [2]

Microbes make our food much more interesting. Imagine a diet without foods made using micro-organisms. I'm not so sure about genetically modified foods though.

Back in Brazil we make our own fuel for cars from sugar cane wastes. This is environmentally friendly and we do not depend so much on fossil fuels. Other countries should try to find alternative sources of energy.

I'm a diabetic. Without micro-organisms to make my insulin, I'd have to use insulin prepared from animals. Genetic engineering has improved the treatment of my diabetes because I use human insulin.

- We all started life as a fertilised egg, just visible to the naked eye. This module looks at the invisible world of microscopic organisms. Micro-organisms are so small that millions of them can fit into a cubic millimetre.

- Micro-organisms are immensely important in our everyday lives. Some of them cause disease, but many of them do useful jobs and form the basis of multi-million pound industries. Micro-organisms make enzymes, foods, fuels and medical products. They improve the fertility of soils and can also clean up the environment. Food chains in the oceans depend on microscopic life.

- Scientists have learned how to change some of these organisms to make them even more useful. Techniques developed from these experiments are now being used to modify plants and animals using the new science of genetic engineering.

What you need to know

- Some micro-organisms are harmful and cause disease, while there are others, like yeast, which have beneficial uses.

- Micro-organisms are involved in the decay of organisms and in the recycling of nutrients.

- Scientists are able to change the characteristics of organisms by selective breeding and genetic engineering.

Microbes by the million

In this item you will find out

- about the structure of bacterial cells and their different shapes
- how bacteria reproduce
- about the production of yoghurt

▲ Foods that use bacteria in their manufacture

Can you imagine the whole of the population of Europe in a raindrop? Or the population of the world living on your skin? This gives you some idea of the size and numbers of **bacteria** in your surroundings.

Bacteria come in a variety of shapes. Some are rod-shaped (bacilli), some are spheres (cocci), some are spiral-shaped (spirilla) and some are curved rods.

How can something so small play such an important part in our daily lives? Many bacteria are useful. They play a vital role in decay and the recycling of nutrients. Billions of bacteria live in your gut and help to keep you healthy. Bacteria are used in the manufacture of foods such as cheese and yoghurt. Genetically modified bacteria make insulin, enzymes and medicines. Other bacteria are harmful and cause serious infections in plants and animals.

Bacteria-like organisms were probably the first forms of life on Earth. Some bacteria make their own food using a primitive form of photosynthesis while others feed on chemicals. But most of them feed on organic nutrients by breaking down organic matter. This means they can survive on a wide range of energy sources and in a wide variety of habitats. They can live in ice, the deepest parts of the oceans, in volcanic springs, embedded in rocks, in acid conditions that corrode metals, in the total absence of oxygen, and even inside us!

▲ The two most common shapes of bacteria are spheres (cocci) and rods (bacilli)

 a Suggest why their incredibly small size makes microbes so difficult to control in places like hospitals or food preparation areas.

Amazing fact

Bacteria are by far the most numerous living organisms on Earth. They are the most ancient forms of life on Earth and have been around for about four billion years.

Bacterial cells

So what is a bacterial cell like?

b Measure the length (in micrometres) of the drawing of the bacterium in the diagram (1 mm = 1000 micrometres).

(i) Calculate the magnification of the diagram (magnification = size of diagram divided by actual size of bacterium).

(ii) A kitten is 0.5 m from its nose to the end of its tail. If this kitten was magnified by the same amount as the bacterium, how long would it be? Give your answer in kilometres.

The cell wall maintains the shape of the bacteria cell and stops it from bursting. The DNA controls the activities of the bacterial cell and how it reproduces (replicates). A bacterial cell is able to move because of its **flagellum**.

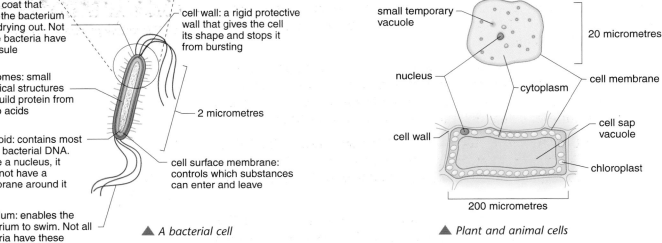

capsule: a sticky coat that stops the bacterium from drying out. Not all the bacteria have a capsule

ribosomes: small spherical structures that build protein from amino acids

nucleoid: contains most of the bacterial DNA. Unlike a nucleus, it does not have a membrane around it

flagellum: enables the bacterium to swim. Not all bacteria have these

cell wall: a rigid protective wall that gives the cell its shape and stops it from bursting

2 micrometres

cell surface membrane: controls which substances can enter and leave

▲ A bacterial cell

small temporary vacuole

nucleus

cell wall

20 micrometres

cell membrane

cytoplasm

cell sap vacuole

chloroplast

200 micrometres

▲ Plant and animal cells

Bacterial cells are very different to plant or animal cells. Bacterial cells do not have a proper nucleus and they also do not contain mitochondria, chloroplasts or a vacuole.

c Write down two other ways a bacterial cell is different to plant and animal cells.

How bacteria reproduce

Bacteria can reproduce by dividing into two. This is called **binary fission** and is a type of asexual reproduction. One bacterium, if left in ideal conditions for 20 hours, has the potential to produce a population of 50 000 000 000 000 000 000 000!

Because bacteria reproduce so rapidly, they can be grown in large tanks called **fermenters**. Fermenters can be used to produce medicines, enzymes, fuels and even new types of food.

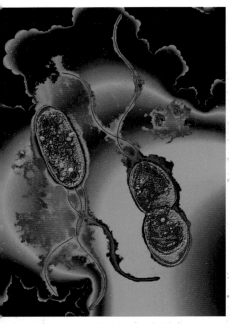

▲ A bacterium reproducing by binary fission

E. coli is a rod shaped bacterium that lives in the gut of humans and other animals, and it can divide into two every 20 minutes. Food can become contaminated with *E. coli* when food handlers do not wash their hands thoroughly after going to the toilet or if they allow raw food to come into contact with cooked food. Flies, contaminated dishcloths and dirty work surfaces soon spread bacteria around a kitchen and onto food. If you eat food that contains *E. coli* then you can suffer from severe food poisoning. You can kill *E. coli* and most other bacteria by heating the food to above 65°C.

 Suggest what precautions you should take to prevent the contamination of food by bacteria.

Making yoghurt

The potential of bacteria to reproduce rapidly in food has led to some accidental discoveries. The production of vinegar, cheese and yoghurt all depend on bacterial activity. Today's food industry is based on a scientific understanding of this process.

Yoghurt is made industrially by batch fermentation, several hundreds of litres at a time. The equipment for making yoghurt is made of stainless steel. Stainless steel is used because it can be **sterilised** using either bleach or steam (steam is non-polluting and leaves no poisonous residue).

After the receiving tank has been sterilised it is filled with raw milk. This is milk taken directly from a cow. Raw milk is heated to 72°C for 15 seconds and then rapidly cooled. This is called **pasteurisation** and kills any active bacteria in the milk.

The milk is then warmed to 40°C before adding two bacteria (Lactobacillus and Streptococcus), which turn the milk sour. These live bacteria are called a **culture**. These two bacteria ferment lactose (the sugar in milk) and turn it into lactic acid. The fall in pH makes the milk curdle and gives the yoghurt its characteristic sharpness.

The yoghurt producer takes samples of the milk and yoghurt at different stages and checks the populations of bacteria. When the producer is happy with the consistency and flavour, the yoghurt is cooled. The producer then may add different fruit flavours, sweeteners or colours before packaging the yoghurt for sale.

 Suggest why it is necessary to sterilise the equipment before making a fresh batch of yoghurt.

 Raw milk may contain a number of different bacteria. Suggest where these bacteria could have come from.

 Suggest why it is important to check the bacterial population at different stages in the production of yoghurt.

▲ *This factory produces yoghurt*

Amazing fact

Some yoghurts contain beneficial bacteria that prevent the growth of other harmful micro-organisms in your gut.

Keywords

bacterium • binary fission • culture • fermenter • flagellum • pasteurisation • sterilised

Handling bacteria safely

▶ *Growing bacteria*

Alison works for a small cheese-producing factory. She is researching the effects of different bacteria on cheese by growing bacterial cultures in her laboratory. She knows that all bacteria can be dangerous so she needs to take precautions when she works with them.

First, Alison washes her hands with antibacterial soap and covers a cut with a clean plaster. She then washes down the bench she is working on with disinfectant. All windows and doors to the laboratory are closed and she puts up a sign warning people not to enter while she is working.

The cultures are grown in Petri dishes containing a sterile jelly called agar. The sample of the bacteria is placed onto the agar using a sterilised inoculating loop. Alison seals the lid of the Petri dish and labels it. She never opens a Petri dish after it has been sealed. After all the Petri dishes are sealed and ready, Alison puts them in an incubator. She then washes down the bench with disinfectant again and washes her hands.

After several hours in an incubator, colonies of bacteria have grown and Alison can study them. When she has finished, she doesn't throw away the bacteria, but she immerses the whole Petri dish and the bacteria in a strong disinfectant solution to kill the bacteria and sterilise the Petri dish.

Questions

1 Why does Alison need to take precautions when handling bacteria?

2 Suggest why it is important to wash the bench with disinfectant before attempting to grow bacteria.

3 Why do you think the Petri dishes need to be labelled?

4 Alison also wears a lab coat and covers her hair when she is working in the laboratory. Suggest why.

The war against disease

In this item you will find out

- about the work of scientists in tackling disease

- how bacteria cause disease

- about the stages in the development of an infectious disease

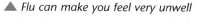
▲ Flu can make you feel very unwell

When did you last have an infection or a disease? What were the symptoms? How did you feel and how long did it last?

For centuries, scientists have been trying to find out what causes diseases and how to treat them. In 1854, Louis Pasteur showed that micro-organisms turned stored wine into vinegar. Pasteur reasoned that if wine was heated before it was stored, then this might kill any living micro-organisms and preserve it. He also found that micro-organisms caused a disease in silkworms. If the micro-organisms were passed on from one silkworm to another, then so was the disease. Pasteur called this the **germ theory of disease**.

In 1865, a young doctor named Joseph Lister introduced the use of carbolic acid disinfectant to prevent fever and infection during surgical operations. Carbolic acid is an **antiseptic**. Antiseptics are chemicals that are used to kill micro-organisms on surgical instruments or on the skin and prevent infection.

▲ Louis Pasteur

▲ Joseph Lister

▲ Alexander Fleming

In 1928, Alexander Fleming noticed that a mould called Penicillium had destroyed neighbouring colonies of bacteria. Ten years later, Howard Florey and Ernst Chain succeeded in extracting the first **antibiotic**, penicillin, from the mould. Antibiotics are drugs that are swallowed or injected into your body. They kill or slow down the growth of harmful bacteria without harming the patient's cells.

a What are the differences between antiseptics and antibiotics?

Amazing fact

The first sample of penicillin was grown in bed-pans and cost thousands of pounds per treatment. Today it is grown in giant fermenters at a cost of just a few pence per treatment.

Feeling unwell?

Some bacteria are pathogens. This means they can cause disease. Diseases can also be caused by viruses and by **protozoa** (single-celled organisms). The table shows five different diseases, what causes them and how they are spread.

Disease	What causes the disease	How the disease is spread
Food poisoning	Salmonella, *E. coli* (rod-shaped bacteria)	Eating raw or undercooked food contaminated with faeces; poor personal hygiene
Cholera	Vibrio (comma-shaped bacteria)	Drinking water contaminated with faeces
Dysentery	Entamoeba (single celled protozoan)	Drinking water contaminated by faeces
Influenza	Influenza (RNA) virus (various types)	Tiny drops of mucus and saliva containing the virus are spread either by coughs and sneezes or by touch
Septicaemia	Staphylococcus (spherical bacteria)	Infected puncture wounds, animal bites or by sharing hypodermic needles

▲ Entamoeba

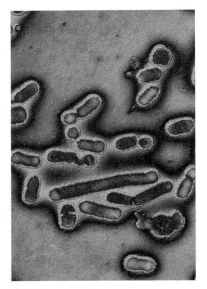

▲ Influenza virus

Spreading disease

▲ These floods in New Orleans were due to a hurricane

Natural disasters, such as earthquakes or flooding, can cause massive amounts of damage to property and can kill thousands of people. Those who survive a natural disaster have an increased risk of catching diseases, such as cholera, food poisoning or dysentery.

Damaged water supplies and sewerage systems mean that drinking water can become contaminated with sewage and disease-causing bacteria. When electrical supplies are damaged, freezers and refrigerators stop working and the food stored in them decays rapidly. If the food cannot be cooked properly, then the harmful bacteria on it will not be killed before it is eaten. Also, if clinics and hospitals are damaged, then people will not have access to health services to treat diseases. This can lead to diseases being spread from person to person.

b Explain how natural disasters can cause diseases to spread rapidly.

Amazing fact

Viruses are so small that it would be possible to pack about 1000 of them into the average bacterium. 500 billion would fit into a cubic millimetre (this is approximately 100 times the world's population).

Life of a disease

What happens when you catch a disease? A sore throat can be caused by a bacterium called *Streptococcus pyogenes*. If a glass has not been properly washed, the rim of the glass may be contaminated with the bacterium. You can catch the disease by drinking from the glass.

Once the bacterium enters your body it multiplies rapidly. During this time, called the **incubation period**, you will not notice any symptoms. This is because the numbers of bacteria are still relatively small and toxins have not yet been released. There is no swelling because the immune system has only just started to respond. The number of bacteria then increases rapidly and they start releasing toxins.

After this, you will find swallowing painful. The lymph glands in your neck begin to swell and your throat looks red and sore and has white patches on it. You will also have a fever. These symptoms appear because the toxins that *Streptococcus pyogenes* releases affect red blood cells and cause pus-forming infections.

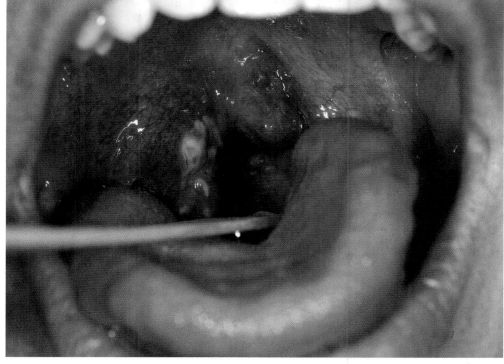

▲ A 'strep' throat

c   Why do the early stages of the infection not produce any painful symptoms?

Water that kills

▲ *Water being supplied by a bore-well*

Danny is a 'barefoot' doctor working in Malawi. He used to work in a hospital in one of the main towns, but for several years he has run clinics in a number of remote villages. The villagers are often very poor and Danny has to treat a lot of different diseases.

Danny is part of a World Health Organisation team trying to educate communities about the importance of clean water supplies for drinking and washing. This team also helps villagers to construct bore-wells to provide clean water. Deep wells are much less likely to be contaminated by sewage. They also lay drains to remove sewage and improve sanitation. The drains ensure that contaminated water does not leak into the villagers' water supply.

Danny has collected data about two waterborne diseases that cause diarrhoea, these are cholera and dysentery. He has also collected data about two other diseases; food poisoning and septicaemia. Food poisoning is caused when food is contaminated by faeces or if it has been poorly prepared or stored. Septicaemia is often caused when insect bites, cuts or scratches become infected.

The graphs show the data for the number of villages with improved drains and water supplies and the numbers of disease cases Danny treated over a period of 10 years.

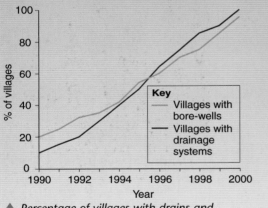

▲ *Percentage of villages with drains and bore wells*

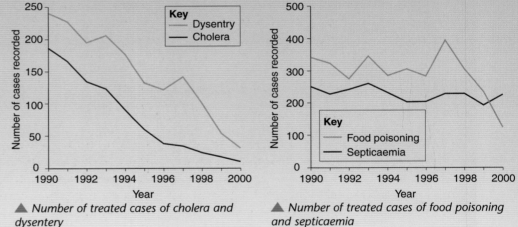

▲ *Number of treated cases of cholera and dysentery*

▲ *Number of treated cases of food poisoning and septicaemia*

Questions

1   By which year did approximately half of the villages have safe water supplies and drainage systems installed?

2   What effect did this have on the number of cases of each of the four diseases?

3   Which two of the four diseases decreased the most with the introduction of safe water supplies?

4   Suggest why these two diseases were not completely eliminated by the introduction of safe water supplies.

5   The summers of 1993 and 1997 were particularly hot. What effect did this have on the number of cases of food poisoning? Suggest a reason for this.

6   Suggest why the number of cases of septicaemia remained fairly constant over the 10 year period.

What's brewing?

In this item you will find out

- about yeast and the conditions that affect the growth of yeast
- about brewing beers and wines
- how distillation increases alcohol content

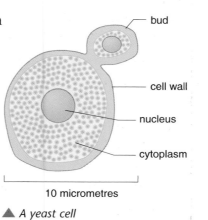

▲ *Yeast helps us to make bread and beer*

Imagine a factory with several thousand workers in every cubic millimetre. Imagine a workforce that never goes on strike and that produces a reliable product 24 hours a day in return only for its food supply. This is a culture of **yeast**!

Yeast is a fungus and it can be used to manufacture fuels, enzymes, vitamins and high quality protein foods. Yeast has been used for thousands of years to make bread and **alcohol** but its potential has yet to be exploited to the full. The photograph shows a colony of yeast.

a What were the two original uses humans had for yeast?

b What are three additional modern uses for yeast?

Yeast can carry out aerobic respiration or anaerobic respiration. During anaerobic respiration (which is carried out in the absence of oxygen), yeast breaks down sugar into alcohol and carbon dioxide. This is called fermentation and means that yeast can turn almost anything that contains sugar into alcohol. The equations below show what happens:

glucose → ethanol (alcohol) + carbon dioxide

$$C_6H_{12}O_6 \rightarrow 2C_2H_5OH + 2CO_2$$

Yeast can also be used to clean waste water from food-processing factories. Water contaminated with sugars is passed over yeast cells that feed on the sugars leaving the water clean.

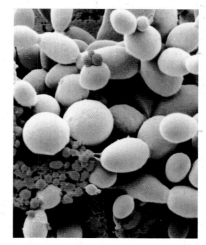

▲ *Scanning electronmicrograph of a colony of yeast*

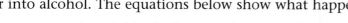

bud

cell wall

nucleus

cytoplasm

10 micrometres

▲ *A yeast cell*

Amazing fact

Yeasts can also be put in fuel cells to turn sugar directly into electricity. Fuel cells containing yeasts may replace batteries to run computers and i-pods.

▲ *Toddy-tree and toddy tapper*

Amazing fact

Elephants sometimes eat over-ripe fruits that have colonies of wild yeasts on their skins. The mixture ferments in the elephants' stomachs releasing large amounts of alcohol that makes the elephants drunk.

Controlling yeast growth

Yeast can reproduce extremely quickly. The optimum growth rate depends on how much food is available for the yeast to use, the temperature and pH of the food and the surroundings, and how quickly the products are removed as they are produced.

Cheers!

People have been brewing wine and beer for centuries. Brewing was probably originally used as a method of preserving fruit juice because the alcohol slows down the growth of other organisms that might spoil the juice. Nowadays, wine is made from grapes, beer from barley, cider from apples, 'toddy' from the sap of palm trees and saki from rice.

Beers and wines are made in several stages. Beer is made from barley. Barley seeds store starch. Yeast cannot ferment starch, so the barley is first allowed to germinate to produce malt. The second stage, called mashing, uses hot water to extract the sugar from the malted barley. Hops can be added as flavouring and the caramel from roasted malt can be added as colouring to make light or dark ales.

The third stage is fermentation which begins when yeast is added to the sugar solution. Fermentation takes place in tanks. Specific yeasts thrive at particular temperatures, usually this is between 15°C and 25°C, but some high alcohol lagers are brewed at lower temperatures. This produces a crisp, light product sometimes called 'ice beer'.

Brewing at lower temperatures saves energy but takes longer. The carbon dioxide produced during fermentation forms a protective layer over the surface of the liquid that helps to prevent other micro-organisms getting in and turning the alcohol into vinegar.

 c Look at the graph. What is the final alcohol content of the lager beer being brewed?

d Suggest one advantage and one disadvantage of brewing lager at 21°C instead of at 11°C.

When most of the sugar has turned to alcohol, the beer has to be 'cleared' to get rid of any sediment. Commercial brewers filter the beer to clear it before it is bottled. Beer is also pasteurised before bottling to kill off any remaining yeast after the main fermentation is finished. This allows the brewers to add extra sugars, chemicals and sweeteners to the beer without worrying that further fermentation will take place within the bottles, which could cause the bottles to burst. This is particularly important when beers are bottled in glass bottles.

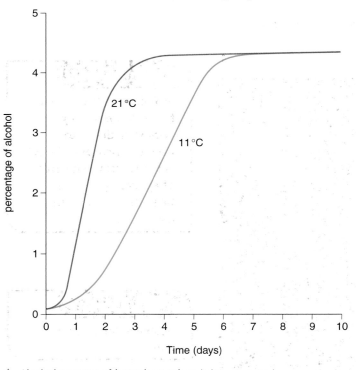

▲ *Alcohol content of lager beer when it is brewed at different temperatures*

Throughout the brewing process, the pH, temperature and sugar content of the beer is monitored. Suggest why.

Stopping fermentation

During fermentation, the alcohol produced by the yeast stops the cells dividing. Eventually the concentration of alcohol kills the yeast. All the yeast is killed off when the alcohol reaches a certain level. Different types of yeast can survive in different levels of alcohol up to about 12%. Wine and beer have different alcohol levels so you use different yeasts for making wine than for making beer.

Distillation

If you want to increase the alcohol level then you have to carry out **distillation**. This is where pure alcohol is separated from fermented sugar by boiling it and condensing the vapour. Spirits, such as whisky, rum and vodka are all prepared in a distillery. In malt whisky, the original sugar comes from barley. In rum the source of sugar is molasses (from sugar cane) and in vodka the sugar comes from digested potato starch.

You can make wine and beer at home, but distillation of spirits has to be carried out in licensed premises. It is illegal to distil alcohol at home.

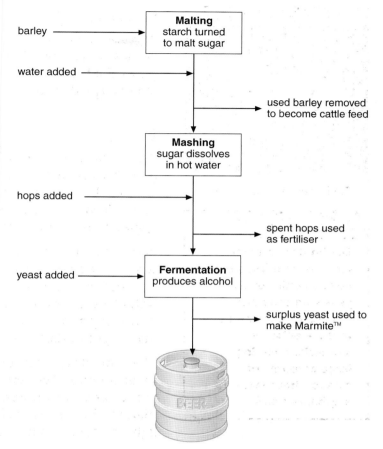

▲ Beer production

▲ Inside a whisky distillery

Amazing fact

In Belgium, they make flavoured beers by adding fruits in the brewing process. 'Kriek' is made using cherries. They even produce beers flavoured with chocolate!

Keywords

alcohol • distillation • yeast

Producing the perfect pint

▲ *Jane is counting yeast cells*

The 'art' of brewing is very much a science. It involves the selection of a suitable variety of yeast and providing it with the precise conditions it needs to ferment the sugar from the malted barley.

Different beers contain different types and amounts of hops selected by the master brewer. Hops affect the taste and flavour of the beer. The amount of sugar in the malt determines the final strength of the beer. Finally, the choice of yeast and the conditions during fermentation are important. Lager beers are often brewed at lower temperatures using specially bred yeast varieties.

Jane Fuggles is a scientist evaluating a new variety of yeast by investigating how temperature affects its growth. She does this by taking samples at regular intervals from yeast cultures kept at different temperatures. She then measures the growth of the yeast by putting a tiny, measured drop of each sample under a microscope. Jane counts the number of yeast cells in a known volume of liquid and compares them. Jane then uses her results to produce a series of graphs.

Jane's graphs show that the yeast growth rate is affected by temperature. In ideal conditions, between 10°C and 40°C, the yeast's growth rate doubles for every 10°C rise in temperature. Above 40°C yeast cells slow down their rate of growth and begin to die.

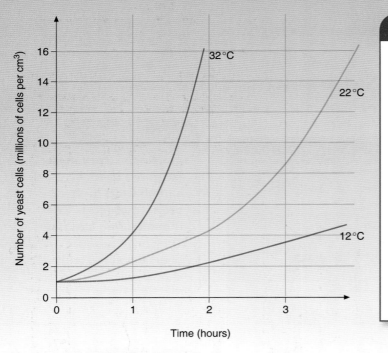

▲ *How temperature affects yeast growth*

Questions

1 How long did it take for the yeast population to increase from 2 million cells per cm^3 to 4 million cells per cm^3:
(a) at 12°C
(b) at 22°C
(c) at 32°C?

2 Do the results of Jane's experimental graphs agree with the statement that the growth rate of yeast doubles with every 10°C rise in temperature? Explain your answer.

3 How long would it take for the number of yeast cells to reach 16 million per cm^3 at 12°C?

4 Copy Jane's graph and add a line to show Jane's expected results at 52°C.

Green energy

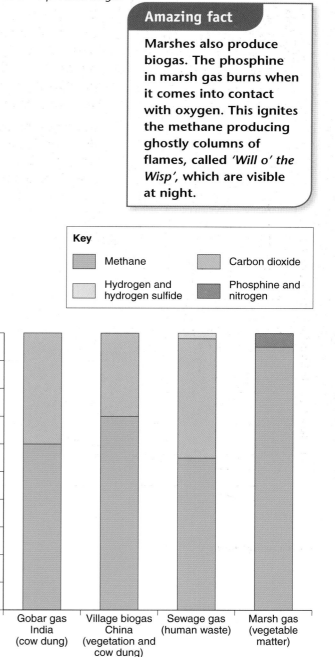

▲ Cows produce biogas

In this item you will find out

- about the composition of biogas and how it is produced

- about the uses of biogas and its advantages as a fuel

- about other biofuels and their effects on the environment

Amazing fact

Marshes also produce biogas. The phosphine in marsh gas burns when it comes into contact with oxygen. This ignites the methane producing ghostly columns of flames, called 'Will o' the Wisp', which are visible at night.

Every day you and millions of other people flush some very important material down the toilet. In the septic tank or at the other end of the sewer are billions of hungry bacteria eagerly waiting for their next meal. Meanwhile another group of bacteria are busy breaking down organic material in the compost heap. Down on the farm, Daisy the cow is breaking wind due to the bacteria living in her gut. What is the link between these three events? The answer is they all generate a mixture of gases called **biogas**.

Biogas is the name given to any flammable gas of biological origin. It is usually a mixture of **methane**, carbon dioxide, hydrogen, nitrogen and hydrogen sulfide (the smelly part). Depending on which organic matter is being broken down, the composition of biogas will vary. Biogas with a methane content above 50% burns easily. Biogas with a methane content of about 10% can be explosive.

a Which group of organisms is responsible for the production of biogas from organic material?

Hydrogen sulfide (which smells like rotten eggs) comes from the breakdown of protein. All of the other gases in biogas are odourless. Foods rich in sulfur, such as beans, onions, eggs and cabbage, have a notorious effect on the composition of biogas. What is more, hydrogen sulfide stops the methane-producing (methanogenic) bacteria from working so well.

Key

- Methane
- Carbon dioxide
- Hydrogen and hydrogen sulfide
- Phosphine and nitrogen

▲ The percentage composition of four types of biogas

195

▲ *Electricity is generated from biogas at this landfill site in Suffolk*

How biogas is formed

Animal dung, vegetable peelings and compost all generate biogas. Anaerobic bacteria first turn the organic material into methanol, hydrogen and a mixture of formic and ethanoic acids. Methanogenic bacteria use these compounds to make methane and carbon dioxide. This is biogas. To do this efficiently, oxygen must be excluded and the temperature must not fall below 15°C. If oxygen is present, most of the organic matter is turned into carbon dioxide and water. It is important to add organic matter and keep the mixture moist. Allowing air into a biogas generator reduces the amount of biogas produced and can produce explosive mixtures of gases. The production of biogas varies so methods have to be used to store it safely.

b Explain why the presence of oxygen makes the production of biogas less efficient.

Using biogas

Biogas is burned to produce electricity. It can also be burned to provide hot water for central heating systems. Dried animal dung used to be burned as a fuel, but turning it into biogas releases six times more energy than burning it. An added bonus is that it also produces valuable fertiliser.

Buses can use biogas as a **biofuel**. Biogas is a cleaner fuel than petrol or diesel, but it releases much less energy when it is burned.

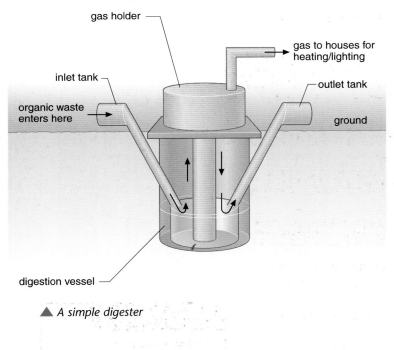

▲ *A simple digester*

Generating biogas commercially

India leads the world in biogas research. The diagram shows a biodigester for making methane on a large scale using a continuous flow process.

Organic material, such as human and animal waste, is regularly tipped into the digester. The digester is sunk into the ground to maintain an even temperature. At low temperatures bacteria reproduce slowly and gas production is reduced. At high temperatures, enzymes are denatured and the bacteria die. The best temperature for biogas production in a village biodigester is between 35 and 45°C. Gas production stops above 60°C.

Biogas collects in a storage tank at the top of the digester. Pathogenic bacteria cannot survive in the biogas generator. The sludge that remains is harmless and can be used as a fertiliser that slowly releases nutrients into the soil.

China has several million village biodigesters. In 1985, biogas stoves and biodigesters supplied up to 40% of the villages' energy. The target is to reach 80%. In Los Angeles, a sewage plant uses biogas from human waste to produce 160 megawatts of power.

c Why does biogas production stop when the digester temperature exceeds 60°C?

 d Why are biogas digesters usually buried below ground?

Biofuels to the rescue

There are many advantages in using biofuels to generate electricity and run vehicles. Biofuels are carbon neutral and cleaner than fossil fuels, such as coal, oil and natural gas. Unlike fossil fuels, when they are burned, biofuels do not add extra carbon dioxide to the atmosphere and produce fewer harmful particulates.

Fossil fuels are limited and they may run out in the future. Carbon dioxide and other greenhouse gases released into the atmosphere when fossil fuels are burned retain heat and act like a blanket round the Earth causing global warming.

Plants harness light energy turning carbon dioxide into sugar, starch and cellulose during photosynthesis. Plants such as oilseed rape produce oils. Growing crops for energy makes sense because burning them only replaces the carbon dioxide they used for growth and does not add to global warming. They are also sustainable resources, which you learned about in B2h.

e Suggest why biofuels are unlikely to supply all the world's energy needs.

Poo power!

Poo is not to be sniffed at! The power of poo is quite astonishing as the information in the table shows.

Type of animal	Amount of dung produced kg/day	Appliance	Amount of dung needed kg/day
Chicken	0.15	Gas fridge for 24 hours	4.5
Cow	23.6	Gas mantle for lighting	1.5
Horse	16.4	Gas ring for cooking meals	5
Human	0.25	Heating water for a bath	1
Pig	3.4	1 kilowatt motor for power	40
Sheep	1.4	An evening of TV	1.2

Three London sewage works use the waste of 6 million people to produce the biogas equivalent to a quarter of a million gallons of petrol per day.

 f i How many tonnes of human dung are produced each day in London (1 tonne = 1000 kg)?

ii How many baths of water would this amount of dung heat?

g How many chickens would produce enough dung to run a fridge constantly for 24 hours?

Biofuels: sweet success for Brazil

Brazil is the world's biggest producer of sugar. Growing sugarcane is becoming increasingly important as a renewable source of energy. Growing sugarcane needs high temperatures, an abundant water supply and extensive land resources. Fortunately Brazil has all of these. Growing sugarcane also requires relatively little labour, but there is a downside – clearing areas to grow sugarcane is leading to a loss of biodiversity.

After harvesting, the sugarcane is milled and pressed. This produces a sugary juice, garapa, and a fibrous residue called bagasse. Garapa is fermented to make alcohol and bagasse is burned to generate electricity

▲ *A sugarcane mill*　　　　　　　　　　　　　　▲ *A car filling up on E85 fuel in Brazil*

E85 is a fuel consisting of 85% alcohol and 15% petrol. Brazil makes 12 billion litres of E85 each year, enough to power its 9 million cars.

This is not a global solution. If the whole of Germany was planted with sugarcane, and even if it could grow, it would only provide half of Germany's energy needs. 'Gasohol' containing 90% petrol and 10% alcohol is sold in Europe. This is a step in trying to reduce our dependence on fossil fuels.

Questions

1 What are the advantages and disadvantages of growing sugarcane to make biofuels?

2 How many billion litres of alcohol are turned into E85 fuel? Show your working.

3 Suggest why it would be impossible for Europe to produce all of the E85 it would need.

4 Some energy is needed to produce E85 from sugarcane. Suggest two processes, needed to produce E85, which use energy.

The fertile soil

In this item you will find out

- about the variety of organisms living in the soil and their feeding relationships

- how earthworms affect soil structure and soil fertility

- about the part soil bacteria play in the nitrogen cycle and the recycling of nutrients

▲ Soil provides a habitat for a variety of invertebrates

Life above ground is there for everyone to see. Life below ground is just as diverse and is essential in maintaining the recycling of important elements needed for plant growth.

Without the action of soil life we might have to climb over dead dinosaur bodies to get to school and many important elements would be unavailable.

What grows at an average rate of 1 cm per thousand years? The answer is the depth of soil. This is the average time it takes for rock to erode, although it can vary between 100 years and 2500 years.

Soil is a complex mixture of rock particles, dead organic material, water and air. It also supports a thriving population of living organisms. These organisms are very important. They break down dead organic matter that releases vital nutrients into the soil.

Without these nutrients we would not be able to grow plants for food. If they did not break down dead organic matter we would be surrounded by the bodies of dead animals and plants.

Only 11% (1.5 billion hectares) of the Earth's ice-free land surface has soil that is cultivated. The remaining soil is too dry, too wet, too poor in nutrients, too shallow or too cold. Some experts believe that as much as 24% of the land surface has the potential to grow crops.

 List the reasons why soil organisms are important.

Amazing fact

75 billion tonnes of soil disappears into the ocean every year. Each year 0.25 million tonnes of soil washes into the River Ganges from the mountain slopes of Nepal.

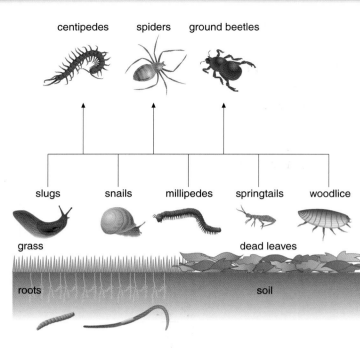

centipedes   spiders   ground beetles

slugs   snails   millipedes   springtails   woodlice

grass

dead leaves

roots

soil

wireworms   earthworms

▲ Soil food web

Life in the soil

Herbivores, such as slugs, snails and wireworms, feed on living plant matter such as leaves and roots. Detritivores, such as earthworms, millipedes and springtails, feed on dead organic matter. **Carnivores**, such as centipedes, spiders and ground beetles, feed on herbivores and detritivores.

b The numbers of herbivores and detritivores greatly exceeds the number of carnivores in the soil food web. Explain why.

c Which group of organisms is likely to be the most abundant in the leaf-litter on the woodland floor? Explain why.

d Wireworms are a troublesome pest causing damage to crops. Farmers construct 'beetle banks' to provide habitats for carnivorous ground beetles and centipedes. How could this help to reduce the wireworm problem?

Improving soil

Most soil organisms and plant roots need water and oxygen to survive. This is because plant roots need water to enable them to take up minerals from the soil. Many soil organisms dehydrate rapidly if the soil dries out. In the summer they burrow deeper down to find moisture. Both plants and animals need oxygen for respiration.

Some heavy soils get waterlogged from flooding, which means that oxygen is kept out. Aerating heavy soil by making holes in or adding compost or gravel allows oxygen to enter and creates passages through which water can drain.

▲ Wireworms are eaten by ground beetles

If a soil is too acidic, then organisms have problems surviving. Most plants are unable to take up minerals from acidic soils. It is important to neutralise acidic soils to improve crop growth. Adding lime to soil neutralises the pH and makes minerals available to the plants. It also makes clay soils easier to work by machines.

Mixing up soil layers stirs up the soil and mixes in extra dead and decaying organic matter (humus) which, when broken down by bacteria and fungi, improves soil fertility.

▲ Soil being limed

e What can be done to improve a heavy soil? Explain how this action improves the soil.

Nature's ploughmen

Earthworms are extremely important organisms for improving the fertility and structure of soil. They bury dead organic matter, which is then broken down by fungi and bacteria. Burrowing aerates and drains the soil by opening up air passages. It also helps to mix up the soil layers.

▲ Earthworms are very useful

The nitrogen cycle

Why is it so important that elements, such as nitrogen, carbon, sulfur and phosphorus, are recycled? Carbon, the basis of all organic molecules, is absorbed from the atmosphere during photosynthesis. Sulfur and phosphorus are both used to make proteins. Without phosphorus plants cannot make DNA, respire or photosynthesise.

Nitrogen is essential for the production of plant protein. Although it makes up 79% of the atmosphere, most plants are unable to use nitrogen directly so they absorb nitrates from the soil instead.

The recycling of these elements depends on different types of bacteria. The diagram shows the bacteria involved in the nitrogen cycle.

The nitrogen cycle involves four types of bacteria:

1 **Saprophytic soil bacteria** decompose dead organic remains and turn them into ammonia.

2 **Nitrifying bacteria** (Nitrosomonas and Nitrobacter) live in the soil and turn ammonia into nitrates.

3 **Nitrogen fixing bacteria** can live in the soil (Azotobacter and Clostridium) or in root nodules (Rhizobium). Root nodules are swellings on the roots of a plant belonging to the pea family. These turn nitrogen gas into nitrates.

4 **Denitrifying bacteria** (Pseudomonas) in waterlogged soil turns nitrate into nitrogen gas.

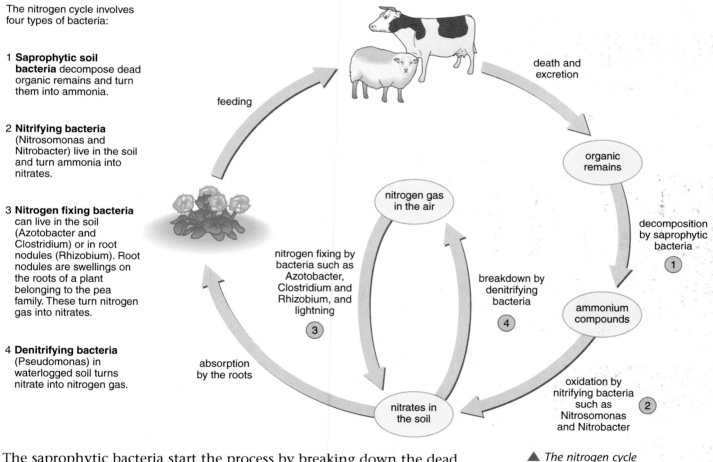

▲ The nitrogen cycle

The saprophytic bacteria start the process by breaking down the dead organic matter. This is the first step in decomposition. Nitrifying bacteria such as Nitrosomonas and Nitrobacter convert ammonia into nitrates. Nitrogen fixing bacteria such as Azobacter, Clostridium and Rhizobium absorb nitrogen from the atmosphere and convert it into nitrates.

 f Farmyard manure contains ammonia. Which group of bacteria are responsible for turning this into nitrates?

 g Parts of the roots of clover plants contain Rhizobium bacteria. Suggest why farmers sometimes plant clover to improve the fertility of soil.

Keywords

carnivore • herbivore

▲ A wormery

▲ Worm casts and plugged burrows

Studying earthworms

Charles Darwin studied earthworms. He was interested in the behaviour of worms and their response to vibrations. He noticed that worms come to the surface when it rains or if you stamp your foot on the ground. He investigated their response to different sound frequencies and constructed a wormery with glass sides so he could observe their behaviour underground. He noticed that worms dragged leaves down into their burrows and realised that this played an important part in improving soil structure and fertility.

Worms pass soil through their intestines, digesting the organic part and releasing casts of fine soil from the anus. Some species of worm leave their casts on the surface (look at a lawn in the autumn). Darwin weighed some of these casts and calculated that worms bring 25 tonnes of soil to the surface per acre each year. In 10 years, the casts would cover a field to a depth of 70 cm. He called worms 'nature's ploughmen'

Worms are found in every continent except Antarctica, and there are many species of worms. Not all of them plug their burrows or leave worm casts, but they all mix up and aerate the soil. Some species of worm are almost 2.5 cm in diameter and others are well over 1 metre long. Their contribution to soil fertility is immense. Gardeners often use wormeries containing tiger worms to turn green waste into rich compost. The compost improves soil texture and returns valuable nutrients to the soil.

Questions

1 The sides of a wormery should be kept covered with black paper. Suggest why.

2 What would you expect to see happening to the layers of soil in a wormery after a few weeks?

3 How could you use a wormery to discover:

(a) the feeding preferences of the worms?

(b) the type of soil the worms prefer to live in?

A watery life

In this item you will find out

- about the advantages and disadvantages of living in water

- that plankton form the basis of many aquatic food webs

- the effects of water pollution on organisms

▲ *An outdoor swimming pool on a summer's day*

Have you noticed how popular a swimming pool is on a hot summer's day? Have you ever imagined what it would like to be a fish and live in water? About 70% of the Earth is covered with water and life in the water is different to life on land. Unfortunately, because there is so much water, humans have used it to get rid of our waste. This has caused a lot of damage to aquatic life.

There are lots of advantages to life in the water. Most organisms need water to survive so being surrounded by water means they will not dehydrate. Large masses of water maintain a much more even temperature than on land. This means that organisms do not have to try to survive in extreme temperatures.

Water also provides support. You can see this if you try weighing an object in air and then weighing it in water. Although it is dense, movement through water is relatively easy because you do not need to support your weight. Waste products from organisms can be released into the water where they will be removed.

Amazing fact

Mayflies spend most of their lives as larvae living in water. As adults they do not feed and live only for a few hours – long enough to mate and lay eggs.

▲ *Dragonfly larvae in water*

▲ *An adult dragonfly*

What do insects, such as dragonflies, and frogs have in common? They both spend some of their life cycle in water and some of it on land. This means they can exploit the benefits of both habitats.

a   **Explain the benefits of living in water.**

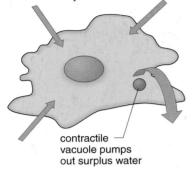

water enters by osmosis

contractile vacuole pumps out surplus water

▲ *An amoeba with contractile vacuole*

▲ *Salmon return to the rivers where they were born to breed*

Problems with water

There are problems with living in water. If you have ever tried to walk in a swimming pool you will know that water provides a greater resistance to movement than air does. This is why lots of fish are streamlined to be able to move through water quickly.

Living organisms also have to maintain a correct water balance within their bodies. This can be difficult to do if you live under water. If the body fluid of an animal is more concentrated than its surroundings, water will enter by osmosis. This extra water then has to be removed. The reverse happens if the surrounding liquid is more concentrated than the body fluid.

Many single celled organisms, for example amoebae, pump surplus water entering the body into a structure called a **contractile vacuole**, which empties when it is full.

b Although freshwater fish do not drink, they produce large amounts of very dilute urine. Suggest why.

From salt water to fresh water

There are also problems for animals moving from salt water to fresh water. Salmon lay their eggs in rivers. Young salmon migrate to the sea where they stay until they mature. Mature salmon return to their birth river to breed and die. When salmon are in the sea, they tend to lose water from their bodies by osmosis. They adjust their body fluid by drinking water and excreting the salt through their gills. They also produce very little urine. When they enter a river, fresh water tends to enter the salmon's body. To lose the excess water they produce lots of dilute urine. They also secrete mucus from their skin and gills so water does not enter their bodies so easily.

c (i) Explain why salmon need to drink when they are living in sea water.
(ii) How does their behaviour change when they enter a river?

Plankton

The Southern Hemisphere is 90% water and is home to the world's largest mammals. Where does their food come from? The answer is **plankton**, small plants and animals that drift on the sea's surface to a depth of less than 100 metres. **Phytoplankton** are microscopic plants and zooplankton are microscopic animals. Phytoplankton make food by photosynthesis.

The size of the plankton population depends on light, temperature and nutrients such as nitrates and phosphates, so seasonal changes can cause changes in the plankton population. Ocean currents circulate warm water but severe storms can pull in cold water from the poles or from deep in the ocean. Scientists are uncertain about the consequences a change in climate might have on plankton populations.

Seasonal changes cause **algal blooms.** These are rapid increases in the amount of phytoplankton brought about by increased light and temperature or nutrients circulated during storms.

The diagram shows a food web in the Antarctic.

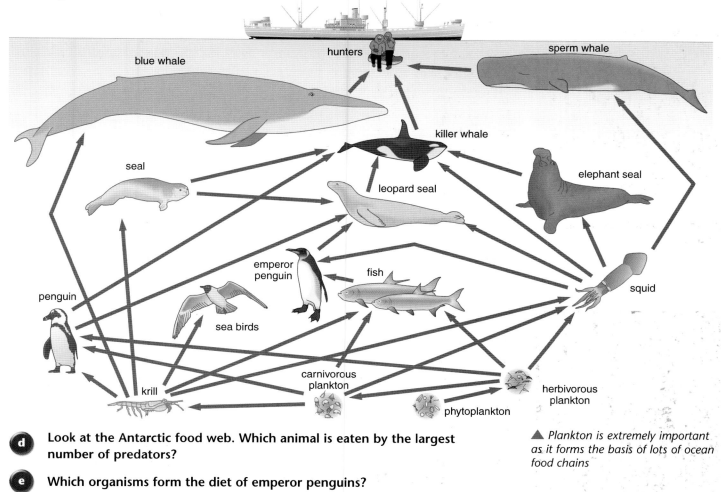

Plankton is extremely important as it forms the basis of lots of ocean food chains

d Look at the Antarctic food web. Which animal is eaten by the largest number of predators?

e Which organisms form the diet of emperor penguins?

f Which organisms are in direct competition with the sperm whale?

g Explain why the food chain that includes the blue whale is more efficient than the food chains involving sperm whales and killer whales.

Water pollution

Streams and rivers can become polluted. Common pollutants are pesticides and fertilisers from farmland and sewage, and chemical spillages from towns and factories. Too many nutrients from sewage or fertiliser cause a rapid growth of algae. This kills plants by blocking out the light. It also kills animals by reducing the amount of dissolved oxygen in the water which means that they cannot respire. This is called **eutrophication**.

Chemicals in the food chain

Environmental scientists have detected two poisons building up in the bodies of marine organisms. DDT (an insecticide) and PCBs (used in electrical equipment) are artificially produced chemicals. They enter the sea in river water. Plankton absorb these chemicals but cannot break them down. DDT and PCBs pass up the food chain becoming concentrated in the fat. Animals at the top of long food chains contain high concentrations of these chemicals, which can cause cancer and liver damage.

Amazing fact

Blooms of plankton are sometimes called 'marine pastures'.

Keywords

algal bloom • contractile vacuole • eutrophication • phytoplankton • plankton

Testing the water

John Bower is a river inspector. His task is to check the water quality of rivers. One way John detects water pollution is by measuring its dissolved oxygen content. He uses an oxygen probe to take readings. He also takes samples of small animals living in the river.

Some animals, such as mayfly and stonefly larvae, are sensitive to the tiniest traces of pollution. They need high levels of dissolved oxygen. Other animals, like freshwater shrimps and water lice, tolerate some pollution and survive in moderate levels of dissolved oxygen. A few animals, like rat-tailed maggots and bloodworms, can live in quite foul water with low levels of dissolved oxygen. All these animals are 'biological indicators'.

If a river supports an abundant variety of life (a high biodiversity) then the water quality is good. If only a few, sluggish, worm-like organisms are present then beware! Similarly, active fish, like trout, need high oxygen levels. They are also sensitive to pH. If water becomes too acidic, trout are unable to survive.

The table shows some of John's data for two rivers.

▲ *A river inspector examining samples of river water*

Type of animal		Number of animals found in each sample of river water	
		River Ocra	River Brignell
Bloodworms		2	12
Flatworms		8	1
Freshwater shrimps		12	3
Mayfly larvae		7	0
Pond snails		5	3
Rat-tailed maggots		0	9
Stonefly larvae		7	0
Water lice		14	4

Questions

1 Which of the two rivers has the best water quality? Give two reasons for your answer.

2 Suggest why active fish, like trout, need water with high levels of dissolved oxygen.

3 Sampling the animals in a river can give a more reliable indication of water quality than taking a few readings with an oxygen probe. Suggest why.

4 Which organism seems most sensitive to pollution, flatworms or pond snails? Give a reason for your choice.

It all comes out in the wash

Most food stains consist of a coloured substance such as tomato ketchup, egg yolk or fruit juice, which either soaks into the fabric or is attached to the surface. Some of these substances are made up of fats, while others are proteins or carbohydrates. Fats, proteins and carbohydrates are all large molecules that do not easily dissolve in water. In order to get the stains out you will need to use a biological washing powder.

Why are biological washing powders good at removing stains? The answer is because biological washing powders contain three classes of enzymes. These are lipases, proteases and amylases.

When enzymes react with fats, proteins and carbohydrates the products of the reactions are fatty acids, amino acids and sugars. These products are all small, soluble molecules so they easily wash off the fabric. The table shows the type of enzyme, the type of stain it acts on and the products that are left.

▲ Have you ever tried to get tomato ketchup stains out of a white shirt?

▲ Some washing powders contain enzymes

Type of enzyme	Type of stain	Products
Lipase	Fats and oils	Fatty acids and glycerol
Protease	Proteins	Amino acids
Amylase	Carbohydrates such as starch	Simple sugars (usually glucose)

 a To remove all types of food stains, biological washing powders must contain a mixture of enzymes. Explain why.

Amazing fact

Many of the first enzymes to be purified came from yeast. The word enzyme means 'in yeast'.

Examiner's tip

> An enzyme's active site is not the same shape as its substrate (the substance it acts on). The shapes are a complementary fit. A substrate fits into the enzyme like a hand in a glove.

Acids and alkalis

Enzymes are biological catalysts that work best at certain pHs. This is because changing the pH alters the shape of an enzyme. The enzymes in biological washing powder work best at about pH 7.5. If the water is too acidic it alters the enzymes' shapes and the biological washing powder does not work so well. Sodium carbonate (washing soda) is sometimes used to soften water but this makes it very alkaline – this too will stop the enzymes from working.

Sweet surprises

How can sugar be made sweeter? Again the answer is enzymes. Surprisingly, not all sugars are equally sweet. One of the most common sugars is **sucrose**, often called cane sugar. Sucrose is an example of a disaccharide, or 'double sugar', made by joining two simpler sugars glucose and **fructose** together. An enzyme, called **sucrase** (or **invertase**) splits sucrose into glucose and fructose by adding a water molecule to it.

$$\underset{\text{sucrose}}{C_{12}H_{22}O_{11}} + \underset{\text{water}}{H_2O} \xrightarrow{\text{sucrase}} \underset{\text{glucose}}{C_6H_{12}O_6} + \underset{\text{fructose}}{C_6H_{12}O_6}$$

Glucose and fructose are much sweeter than sucrose and this makes them useful in the food industry. For example, foods such as low-calorie foods for slimmers can be sweetened without needing to add much sugar.

To make soft-centred chocolates, sucrase is mixed with sucrose to make a stiff 'icing sugar' mixture. Melted chocolate is poured over the mixture and left to set. The enzyme liquefies the sucrose producing the soft centre shown in the photograph.

▲ *A soft-centred chocolate*

 b   Explain why breaking down sucrose is useful for the food industry.

Immobilising enzymes

Enzymes are expensive. Most of them have to be extracted from living tissue and purified before they can be used in the food industry. They are valuable in the food industry because they speed up just one reaction in a complex mixture of chemicals. It is important not to 'lose' the enzyme in the mixture and not to contaminate the food product with it.

Immobilising enzymes means attaching them to something, like plastic beads or a sheet of material. This allows them to be removed from the mixture when the reaction is complete. Another way of immobilising enzymes is to surround them in a permeable jelly, called alginate, which is made from seaweed. These **alginate beads** can be used in continuous flow processing. The beads containing the immobilised enzyme are placed in a container called a reaction vessel. The liquid to be treated flows through the spaces between the beads in the reaction vessel, producing a continuous supply of the treated product for immediate use. This is much more convenient than waiting to treat separate batches of liquid or having to store the treated product until it is needed.

c

(i) Suggest an advantage of immobilising enzymes.

(ii) Why is it better to immobilise enzymes in lots of small alginate beads rather than in fewer, larger beads?

Lactose-free milk

Lactose or 'milk sugar' is present in cows' milk. Some people, and cats, cannot digest lactose. If these people drink milk, the lactose travels through the gut to the large intestine. In the large intestine, bacteria ferment the lactose making lots of gas. The undigested lactose also interferes with the absorption of water by the colon causing diarrhoea. Gas and diarrhoea is not a pleasant combination!

To make lactose-free milk, an enzyme, lactase, is mixed with a solution of sodium alginate. This mixture is dripped into a calcium chloride solution where it forms beads of immobilised enzyme. The beads are washed and put into a container. Milk is passed through the container so it runs between the beads. Lactase in the beads turns the lactose into glucose and **galactose**. Both of these sugars can be absorbed by people and cats. The milk is not contaminated by the enzyme and the enzyme is not lost because it remains in the beads.

d Explain why it is important that none of the immobilised enzyme leaks out of the beads.

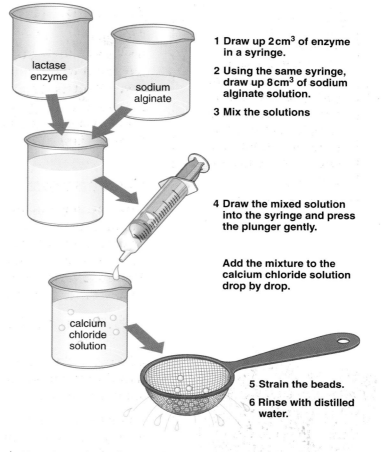

1 Draw up 2 cm³ of enzyme in a syringe.

2 Using the same syringe, draw up 8 cm³ of sodium alginate solution.

3 Mix the solutions

4 Draw the mixed solution into the syringe and press the plunger gently.

Add the mixture to the calcium chloride solution drop by drop.

5 Strain the beads.

6 Rinse with distilled water.

▲ *How enzymes can be immobilised*

▲ *Cat milk doesn't contain lactose*

IMMOBILISING AN ENZYME

Keywords

alginate bead • fructose • galactose • invertase • sucrase • sucrose

▲ Some of the confectionary made by Charley's Chocolate factory

Keeping rivers clean

Charley's Chocolate factory makes sweets and confectionary. Shredded coconut is used to make some of the sweets. Water from the river is used to wash the shredded coconut. Waste water from the factory contains sugar and starch. This must be removed from the water before it returns to the river.

Jenni Ayton advises industry on how to reduce pollution. She suggests two ways of treating the waste water. She says one way is to use immobilised enzymes to break down the starch and sugar. Another way is using micro-organisms that feed on the waste. Micro-organisms are trapped inside pieces of sponge. The waste water is filtered through this sponge. The micro-organisms feed on the starch and sugar leaving the water clean. The food factory can then recycle its own water.

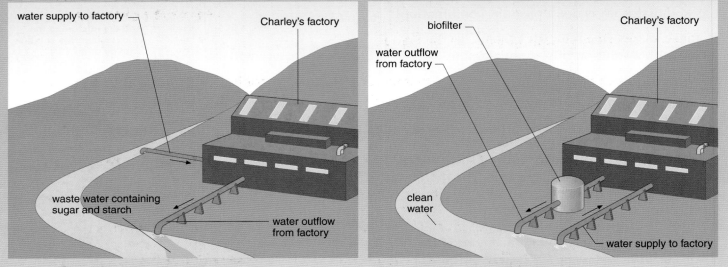

▲ The factory before adding the biofilter   ▲ The factory with the biofilter in place

Jenni recommends this second method because the micro-organisms multiply and produce single cell protein (SCP). The surplus micro-organisms are squeezed out of the sponge and used to make SCP. The SCP is sold to an animal food manufacturer so the treatment pays for itself. If the water intake for the factory is re-sited downstream of the outfall, the factory will then be less likely to allow pollution of the river.

Questions

1 The micro-organisms need to be trapped inside pieces of sponge. Suggest two reasons for this.

2 Write down two advantages of using micro-organisms instead of immobilised enzymes to clean the water.

3 The owner of the factory is less likely to pollute the river if the water supply to the factory is downstream from the waste outflow. Suggest why.

4 If the sugar and starch waste reaches the river it is broken down naturally. This reduces the amount of dissolved oxygen in the river water. Explain why this would affect the organisms living in the river.

Better by design

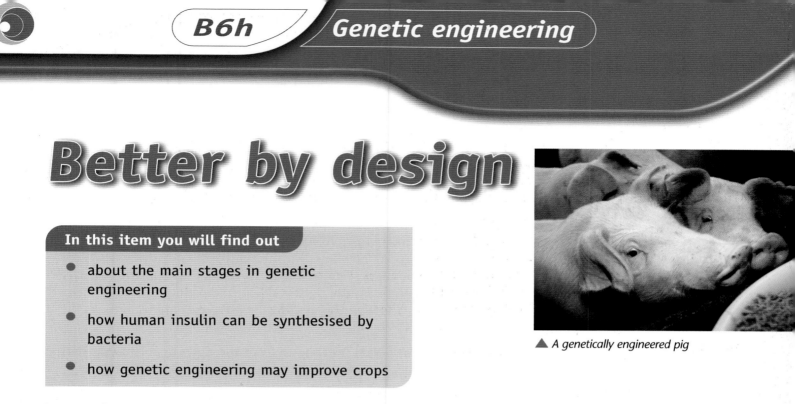

▲ *A genetically engineered pig*

Humans have been producing plants and animals by selective breeding for hundreds of years. Until the mechanism of inheritance was worked out about 100 years ago, this was a very hit and miss process. Even with a modern knowledge and understanding of genetics, this way of producing new organisms has several drawbacks. It is slow, it involves lots of genes and it mixes them at random.

Genetic engineering is quicker, it involves small numbers of genes and it is more predictable. Genetic engineering alters the genetic code of an organism by inserting different genes into it. Organisms produced by genetic engineering are called genetically modified (GM) organisms. It is possible to move genes from one species of organism to another, even from animals to plants or bacteria. The new type of organism containing genes from different species is called a **transgenic organism**.

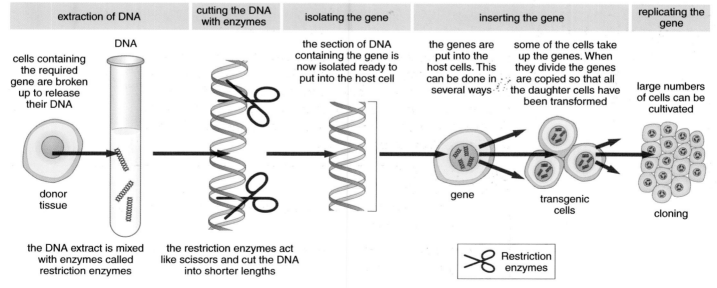

| extraction of DNA | cutting the DNA with enzymes | isolating the gene | inserting the gene | | replicating the gene |

cells containing the required gene are broken up to release their DNA

DNA

donor tissue

the DNA extract is mixed with enzymes called restriction enzymes

the restriction enzymes act like scissors and cut the DNA into shorter lengths

the section of DNA containing the gene is now isolated ready to put into the host cell

the genes are put into the host cells. This can be done in several ways

some of the cells take up the genes. When they divide the genes are copied so that all the daughter cells have been transformed

large numbers of cells can be cultivated

gene

transgenic cells

cloning

Restriction enzymes

▲ *How transgenic cells are produced*

Genetic engineering means that humans can make changes to life on Earth in a very short time. There are lots of advantages in being able to do this, but there are also lots of drawbacks.

a   **Explain how transgenic cells are created.**

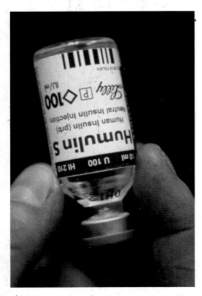
▲ *Human insulin*

Carrying out genetic engineering

So how do you create a transgenic organism? First you have to choose an organism with the desired characteristics. Then you have to find the gene that causes these characteristics and remove the gene from the organism's DNA. You then cut open the DNA in another organism and insert the new gene into its DNA.

A special group of enzymes, called **restriction enzymes**, can cut DNA into short lengths. These enzymes look for particular patterns called sequences in the DNA. When they find the right sequence, they cut the DNA like a pair of chemical scissors. Different restriction enzymes cut DNA in different places. Choosing the right restriction enzymes means you can cut out just the gene you want. Enzymes called **ligases** are used to rejoin DNA strands together.

Once the new gene is spliced into the DNA of the new organism you need to use an **assaying technique** to check that the new gene has been transferred correctly and that it works properly. To be able to check that a gene has been properly transferred, it is first attached to a 'marker gene'. One assaying technique uses a marker gene that produces bioluminescence. This means that it glows in the dark. When the genes and their markers are added to cells, only a few of them are successfully spliced into the cell's DNA. Later, when the cells are examined, the ones that glow in the dark are the ones that have successfully taken up the new gene.

 How do restriction enzymes work?

Once the transgenic organism has been approved then it is cloned to create identical copies.

Making insulin

It is possible to produce transgenic bacteria by using genetic engineering. One type of transgenic bacteria can be used to make human insulin. A marker gene for antibiotic resistance is used to check that the insulin gene has been successfully transferred. The diagram opposite shows how human insulin is created.

The gene that is responsible for producing human insulin is cut out of human DNA using restriction enzymes. Bacteria contain a ring of DNA known as a **plasmid**. These plasmids are cut open using the same restriction enzymes that are used to cut the insulin gene from a human cell. The insulin gene and its marker are then inserted into the bacterial plasmid.

The transgenic bacteria are then cloned and allowed to multiply so the insulin gene gets copied billions of times. Because the genetic code is the same for all living things, the bacteria will make insulin even though they have no use for it. The bacteria become swollen with insulin, which can then be extracted and purified. Large quantities of human insulin can therefore be harvested. People who have diabetes can use human insulin injections to control their blood sugar levels.

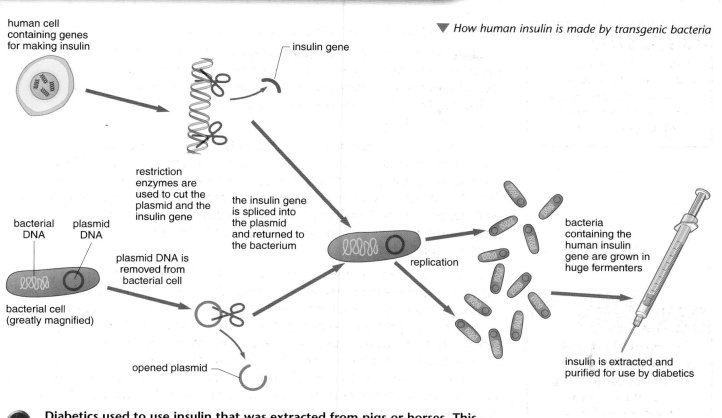

human cell
containing genes
for making insulin

▼ How human insulin is made by transgenic bacteria

insulin gene

restriction
enzymes are
used to cut the
plasmid and the
insulin gene

the insulin gene
is spliced into
the plasmid
and returned to
the bacterium

bacterial
DNA

plasmid
DNA

plasmid DNA is
removed from
bacterial cell

replication

bacteria
containing the
human insulin
gene are grown in
huge fermenters

bacterial cell
(greatly magnified)

opened plasmid

insulin is extracted and
purified for use by diabetics

c Diabetics used to use insulin that was extracted from pigs or horses. This sometimes produced side effects. Human insulin has fewer side effects. Suggest why injecting insulin from pigs or horses might cause more side-effects than human insulin.

Genetically modified crops

Genetic engineering can help the world to produce more food more quickly. It can make plants grow faster and bigger. It can make them resistant to disease and resistant to weedkillers. Soya beans can have a gene inserted making them resistant to herbicides. Weeds in fields of soya can then be controlled without killing the crop. This reduces competition for light, water and nutrients so the soya grows better.

Genetic engineering can also make crop plants grow and survive in poor conditions not usually suitable for growing plants, such as in salty water or in places that suffer from drought.

Finally, plants can be engineered to produce other chemicals, such as vitamins. BTI maize has a gene from a bacterium spliced into its cells. This gene makes a natural insecticide that stops leaf damage by caterpillars. In Asia, many people become blind due to vitamin A deficiency in their rice diet. Genetically modified rice contains the gene to make vitamin A, so people eating it have their sight protected because they no longer develop vitamin A deficiency.

d High doses of some vitamins, such as vitamin A, can be toxic. What advice should be given to people using GM rice?

▲ Genetically modified rice

Keywords

assaying technique
• ligase • plasmid •
restriction enzyme •
transgenic organism

GMOs – the answer to Africa's problems?

Genetic engineering is used to produce genetically modified organisms (GMOs). GMOs include crops that can resist drought, need less fertiliser, make their own pesticides and produce greater yields. Much of Africa suffers from famine. Climate change, over-grazing and loss of vegetation increase erosion and cause desertification (the spread of deserts). Are GMOs the answer to some of Africa's problems?

Genetic modification can quickly improve crop characteristics. It is faster and more precise than selective breeding. Single genes can be added, often from unrelated species. GM crops can produce completely artificial substances such as vaccines or raw materials to make plastic. Genetic modification also enables plants to grow in poor, dry soils holding them together and thus preventing the spread of deserts.

Risks include unexpected allergic responses when the food is eaten. It may result in the spread of pest resistance to wild plants or cause the build up of toxins in food chains. This could cause lasting environmental damage and may affect human health.

African countries have made different conclusions about using GMOs. In Burkina Faso they grow GM cotton that makes its own insecticide to kill the cotton boll weevil. Mali and Kenya are considering growing GM cotton. In South Africa 80% of the cotton and 20–30% of the maize is already GM.

Scientists in Uganda have developed GM bananas but do not have permission to grow them. In Nigeria, growing GM black-eyed beans could increase the yield threefold.

Most other African countries have banned GMOs. Ethiopia says GMOs would undermine farmers who have traditional ways of fighting weeds and pests.

▲ Africa from space

▲ Cotton bollweevil

Questions

1 List the risks and benefits of growing GMOs.

2 Explain why producing GMOs is more precise than selective breeding.

3 Suggest why growing plants that make their own pesticide may cause 'irreversible environmental damage'.

4 How could GMOs be used to delay the effects of desertification?

B6a

1 Look at the diagram of a bacterial cell on page 184. Describe the functions of:

   a the cell wall   **b** the DNA   **c** the flagellum   [all 1]

2 The structure of a bacterial cell differs from the structure of plant and animal cells.

Write down three differences. Use the diagrams on page 184 to help you.   [3]

3 Describe four stages in the preparation of strawberry yoghurt from raw milk.   [4]

4 Bacteria can be harmful, especially in places where food is being prepared. Describe and explain three precautions which should be taken in order to prepare food safely.   [6]

5 In the canning industry, all of the bacteria present in food must be killed.

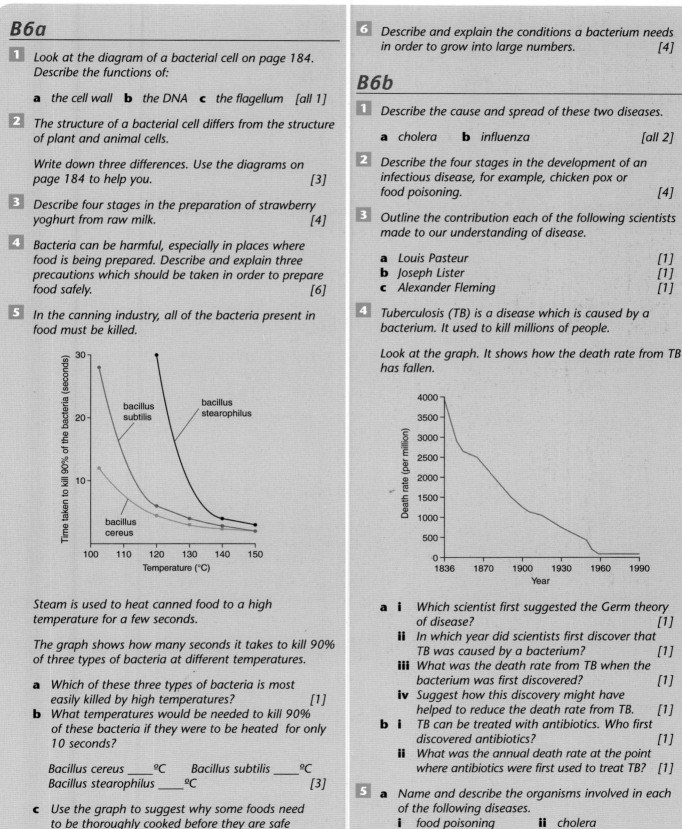

Steam is used to heat canned food to a high temperature for a few seconds.

The graph shows how many seconds it takes to kill 90% of three types of bacteria at different temperatures.

   a Which of these three types of bacteria is most easily killed by high temperatures?   [1]
   b What temperatures would be needed to kill 90% of these bacteria if they were to be heated for only 10 seconds?

     Bacillus cereus ____°C   Bacillus subtilis ____°C
     Bacillus stearophilus ____°C   [3]

   c Use the graph to suggest why some foods need to be thoroughly cooked before they are safe to eat.   [1]

6 Describe and explain the conditions a bacterium needs in order to grow into large numbers.   [4]

B6b

1 Describe the cause and spread of these two diseases.

   a cholera   **b** influenza   [all 2]

2 Describe the four stages in the development of an infectious disease, for example, chicken pox or food poisoning.   [4]

3 Outline the contribution each of the following scientists made to our understanding of disease.

   a Louis Pasteur   [1]
   b Joseph Lister   [1]
   c Alexander Fleming   [1]

4 Tuberculosis (TB) is a disease which is caused by a bacterium. It used to kill millions of people.

Look at the graph. It shows how the death rate from TB has fallen.

   a **i** Which scientist first suggested the Germ theory of disease?   [1]
     ii In which year did scientists first discover that TB was caused by a bacterium?   [1]
     iii What was the death rate from TB when the bacterium was first discovered?   [1]
     iv Suggest how this discovery might have helped to reduce the death rate from TB.   [1]
   b **i** TB can be treated with antibiotics. Who first discovered antibiotics?   [1]
     ii What was the annual death rate at the point where antibiotics were first used to treat TB?   [1]

5 **a** Name and describe the organisms involved in each of the following diseases.
     i food poisoning   **ii** cholera
     iii dysentery.   [6]

b Explain why antibiotics are of no use in the treatment of dysentery. [2]

6 After natural disasters such as floods or earthquakes there is often a rapid spread of disease.

Explain why. [4]

B6c

1 State the word equation for fermentation (anaerobic respiration in yeast). [3]

2 Outline the main stages in the brewing of beer or wine. [5]

3 Rum, whisky and vodka are three drinks made using yeast.
  a Name the source of the sugar used in the production of each of these drinks. [3]
  b Name the process used to increase the alcohol concentration of these drinks. [2]

4 A group of pupils counted the number of CO_2 bubbles produced each minute by a yeast and sugar mixture.

They did their experiment at different temperatures and produced the following graph.

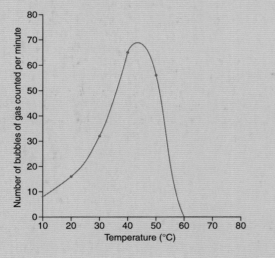

  a i How many bubbles of gas were produced per minute at each of the following temperatures?
    10 °C ____ 20 °C ____ 30 °C ____ [3]
   ii Describe the pattern shown by these results. [2]
  b At which temperature is the rate of production of gas at its greatest?
    Suggest a reason for this. [3]
  c Explain why the production of gas ceased at temperatures above 60 °C. [2]

5 a The final alcohol content of a wine can be increased by changing the conditions during fermentation. Give three ways. [3]
  b Explain why it is not possible to brew a wine with an alcohol content above about 15%. [1]

6 a Explain what is meant by the term 'Pasteurisation' [2]
  b Why is it necessary to do this to bottled beers? [2]

B6d

1 a What is biogas and how is it produced? [2]
  b Give three uses for biogas. [3]

2 Describe the advantages of using biofuels. [3]

3 Explain why it is important to exclude air from a biodigester which produces biogas. [4]

4 Biofuels are a sustainable source of energy and are friendly to the environment. Explain this statement. [3]

5 a What is global warming? [2]
  b Explain why the carbon emissions from the burning of E85 fuel (85% alcohol, 15% petrol) make a smaller contribution to global warming than burning 'gasohol' (10% alcohol, 90% petrol). [3]

B6e

1 The organisms in a soil food web can be put in to three feeding groups.
  a Name these three feeding groups and give one example of an organism in each group. [3]
  b Which of these feeding groups usually has the smallest numbers? Give a reason for this. [2]

2 Earthworms make valuable contributions to improving soil structure and fertility.

Describe four ways in which they do this. [4]

3 Explain why it is important that elements are recycled in the soil. [3]

4 Look at the diagram of the nitrogen cycle on page 201.
  a Name the three groups of bacteria involved in the nitrogen cycle. [3]
  b Describe the parts each of these groups of bacteria play in the nitrogen cycle. [3]

5 Most plants do not grow very well in acid soils. Give a reason for this and state what can be done to correct this problem. [2]

6

a Snails, slugs, caterpillars, beetle larvae and millipedes are all herbivores. Would you expect to find more in a meadow or woodland? Suggest why. [2]

b Worms and woodlice are detritivores. What does this mean? [1]

c Spiders and ground beetles are carnivores. Suggest why there are relatively few carnivores. [1]

B6f

1 List four advantages of living in fresh water. [4]

2

a What name is given to the microscopic plants and animals which drift near the surface of the ocean? [1]

b State two environmental factors which affect the population of these small plants and animals. [2]

3 Sewage and fertiliser can cause eutrophication if it drains into rivers of lakes.

Describe four effects of eutrophication. [4]

4 Animals such as an amoeba which live in fresh water have problems controlling their water balance.

a Name the process by which water enters or leaves animal cells. [1]

b Explain how an amoeba controls the amount of water in its cell. [1]

c What problem does a salmon have to overcome when it moves from sea water to fresh water to breed? [2]

5 Toxic substances such as DDT [an insecticide] have been found in the bodies of marine animals such as whales.

a What term is used to describe this build up of toxic substances? [1]

b Explain how this build up of toxic substances occurs. [4]

6 The diagram shows the concentration of pesticides in an aquatic food chain.

DDT = 80 parts per million	fish eagle
10ppm	needle fish
0.1ppm	small fish (minnows)
0.008ppm	zoo plankton (water fleas)
0.0008 parts per million	phytoplankton (microscopic algae)

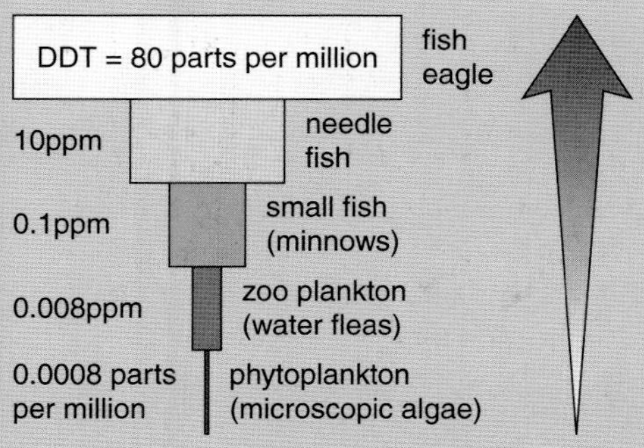

a How many times greater is the pesticide concentration in the fish eagle than in the zooplankton? Show your working. [2]

b Give two reasons why the levels of DDT pesticide increase as you go up the food chain. [2]

c Land-based food chains are usually shorter than aquatic food chains. Suggest why the top carnivore in an aquatic food chain may be more likely to be harmed by DDT pesticide than a land-based carnivore. [2]

B6g

1 Biological washing powders contain three kinds of enzyme. Name these three types of enzyme and state the action of each one. [6]

2 Many automatic washing machines have a special biological washing cycle set at about 40 °C.

State the advantage of this special cycle and explain why a higher temperature would be unsuitable for biological washing powders. [3]

3 The food industry uses an enzyme to break down sucrose into glucose and fructose.

a What is the name of this enzyme? [1]

b Why is this process useful to the food industry? [1]

c How does the food industry prevent the food becoming contaminated with the enzyme? [1]

4

a What are 'immobilised enzymes'? [1]

b State three advantages of using immobilised enzymes. [3]

5 Some people are unable to digest the lactose. This is the sugar found in milk.

a Explain why lactose causes problems for people who are unable to digest it. [2]

b Explain how lactose can safely be removed from milk. [2]

B6h

1 Outline the main stages in genetic engineering. [5]

2 Describe how bacteria can be altered to produce human insulin. [4]

3 State two ways in which crops can be improved by genetic engineering. [2]

4 Name the types of enzymes which are used:

a to cut DNA in specific places [1]

b to rejoin (splice) lengths of DNA together. [1]

5 Using named examples, state three ways in which crop plants have been altered by genetic modification. [3]

6 State three reasons why some people are concerned about the use of genetic modification. [3]

Useful data

Physical quantities and units

Physical quantity	Unit(s)
length	metre (m); kilometre (km); centimetre (cm); millimetre (mm)
mass	kilogram (kg); gram (g); milligram (mg); migrogram (µg)
time	second (s); millisecond (ms)
temperature	degrees Celsius (°C); kelvin (K)
current	ampere (A); milliampere (mA)
voltage	volt (V); millivolt (mV)
area	cm^2; m^2
volume	cm^3; dm^3; m^3; litre (l); millilitre (ml)
density	kg/m^3; g/cm^3
force	newton (N)
speed	m/s; km/h
energy	joule (J); kilojoule (kJ); megajoule (MJ)
power	watt (W); kilowatt (kW); megawatt (MW)
frequency	hertz (Hz); kilohertz (kHz)
gravitational field strength	N/kg
radioactivity	becquerel (Bq)
acceleration	m/s^2; km/h^2

Electrical symbols

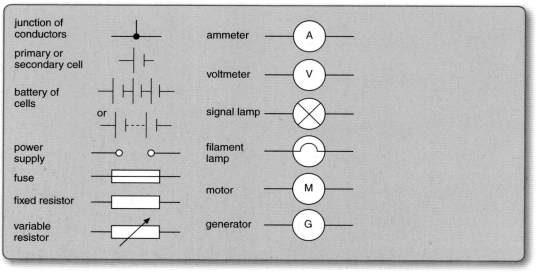

Introduction to Skills assessment

Work carried out by you outside an examination (called 'Skills assessment') is an important aspect of GCSE Biology. This will include practical work, research and report writing – all things a real biologist has to do.

For GCSE Biology you have a choice of the 'Skills assessment' that you submit for assessment.

1 The 'Skills assessment' that is used for Core Science:
'Can-do tasks'
'Science in the news'

2 The 'Skills assessment' that is used for Additional Science:
'Research task'
'Data task'
'Assessment of your practical skills'

Either route is worth one third of the marks available for GCSE Biology. So this practical assessment is very important.

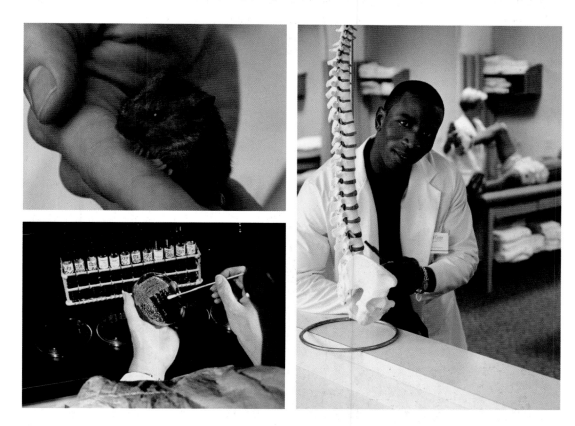

Can-do tasks

Biology is a practical subject and you deserve credit for being able to do practical things. 'Can-do tasks' are an opportunity for you to demonstrate some of your practical and ICT skills throughout the course.

There are 57 of these tasks throughout your GCSE Biology course. Some are practical and some require the use of ICT.

You can only count a maximum of eight of these tasks and these tasks are set at three levels:

Basic (worth 1 mark): These are simple tasks that you can usually complete quickly. You may have done many of these before you started the course.

Intermediate (worth 2 marks): These are slightly harder tasks that might take a little longer to do.

Advanced (worth 3 marks): These are even more difficult tasks that may take you some time to do.

A complete list of Biology 'Can-do tasks' is shown below.

Basic: 1 mark 'Can-do tasks'

1	B1a	I can measure blood pressure.
2	B1a	I can measure breathing rate/pulse rate before and after different types of exercise.
3	B1d	I can measure my field of view.
4	B1d	I can use Ishihara colour charts to identify colour vision deficiency.
5	B1f	I can use ICT to produce a poster warning old people about hypothermia and telling them how to prevent it.
6	B2a, B2b	I can use a simple key to identify some plants/animals.
7	B2b	I can classify some different organisms.
8	B2e	I can use a hand lens to observe a small animal.
9	B2f	I can identify a range of fossils.
10	B2f	I can use the internet to find out information about Charles Darwin.
11	B2h	I can use the internet to collect scientific information about extinct animals.
12	B5a	I can identify a fracture from an X-ray.
13	B5d	I can measure my peak flow.
14	B6c	I can collect gas from fermenting sugar and test it for carbon dioxide.
15	B6f	I can observe a living Daphnia under a microscope.
16	B6g	I can test a mock urine sample for the presence of glucose.
17	B6g	I can use a colour chart to determine how much glucose is in the 'urine' sample.

		Intermediate: 2 marks 'Can-do tasks'			

18	B1a	I can do an experiment on fatigue in finger muscles and record the results.
19	B1b	I can carry out simple food tests.
20	B1b	I can calculate a BMI and make a decision as to what it indicates.
21	B1c	I can collect data from various sources for a named disease and identify danger sites on a world map.
22	B1d	I can collect, present and analyse data to compare the sensitivity of different areas of my skin.
23	B1e	I can collect scientific information from a variety of sources to show the effects of drugs or smoking on the body and display or present the information.
24	B1f	I can carry out an experiment on skin temperatures down an arm or leg and plot the results on a graph.
25	B2a	I can collect data using a sampling technique.
26	B2c	I can measure the rate of photosynthesis by counting the rate of bubble release from pond weed.
27	B2e	I can use ICT to make a poster to explain how a camel/polar bear is adapted to its habitat.
28	B2f	I can use ICT to prepare an information leaflet explaining why the fossil record is incomplete.
29	B2g	I can plot a population graph.
30	B2h	I can use the internet to collect scientific information about various endangered species.

31	B5a	I can identify the main bones and muscles in an arm.
32	B5b	I can construct a time line of discoveries about blood circulation using various sources.
33	B5b	I can display information using charts and graphs about heart disease in the world.
34	B5d	I can carry out an experiment to show the differing amounts of carbon dioxide in inhaled and exhaled air.
35	B5d	I can survey one industrial disease and present the information in a poster or leaflet.
36	B5e	I can investigate urine samples and correctly identify them.
37	B5e	I can demonstrate mouth-to-mouth resuscitation on a dummy.
38	B5h	I can collect and display data to show height distributions in students.
39	B6a	I can follow instructions to produce a sample of yoghurt.
40	B6a	I can measure/record the pH of milk as it is converted to yoghurt using pH paper/ pH meter/data logger.
41	B6e	I can identify some soil fauna and flora using keys.
42	B6e	I can do a simple experiment to show that life is present in a soil sample (using lime water or bicarbonate indicator).
43	B6g	I can immobilise an enzyme in an alginate bead.

Advanced: 3 marks 'Can-do tasks'	
44 B1b I can carry out an experiment on enzyme action and record the results and conclusion.	**51** B6b I can compare the effectiveness of different antiseptics using a culture of bacteria on an agar plate (by measuring and comparing the diameters of the halos).
45 B1h I can use a genetics kit to show a monohybrid cross.	**52** B6c I can make a slide of yeast cells, stain it and make a labelled drawing.
46 B2a I can investigate and compare different habitats.	**53** B6c I can do an experiment to show how yeast activity is affected by temperature.
47 B2b I can present a report on the work of Carl Linnaeus.	**54** B6d I can design a biogas digester and display the plans as a chart.
48 B2c I can test a leaf for starch.	**55** B6e I can compare air content of two different soils.
49 B2h I can use ICT to produce an information leaflet on one endangered species, showing reasons for its predicament and suggestions for its protection.	**56** B6g I can compare the effectiveness of a biological washing powder in removing different stains.
50 B6b I can prepare a culture of bacteria on an agar plate using an aseptic technique.	**57** B6g I can show that my bead contains an enzyme by showing its effect on a substrate.

Some of the 'Can-do tasks' produce a product, for example a leaflet on one endangered species. This can be very useful as evidence that you have successfully completed a task.

Your teacher has to see that you have completed a task and record this on a record sheet. Remember only the best eight can count so the maximum is $8 \times 3 = 24$.

Your teacher may give you a list of all the 'Can-do tasks' at the start of the course. You can then tell him or her when you think you have successfully completed one of them. They can then give you credit for this.

Remember, it does not matter if you fail to do a task or if you are absent. There will be many more chances throughout the course.

Finally, just how important is a 'Can-do task'? Every time you complete an advanced task it is worth more to your final result than scoring 4 marks on a written paper.

Good luck!

Science in the news

Do you read a newspaper or listen to radio or television news programmes?

Do you believe everything you read or hear?

Here are two newspaper headlines:

Bird Flu Man Works in Hospital and **Eating Broccoli Every Week May Stop Lung Cancer**

Looking at the first headline you might think the man had bird flu. But if you had read the article you would have discovered that he was the man who ran a quarantine centre where two parrots died and he also worked in a hospital. He never had bird flu.

Reading the second article would tell you that the sample used was so small that scientists could not be sure that a weekly helping of broccoli would cut the risk of getting cancer.

As part of your GCSE Biology you have to do at least one 'Science in the news task'. If you do more than one, only your best mark will count. In the 'Science in the news task' you have to use your knowledge of science to solve a problem.

The task will be in the form of a question, for example: *'Should smoking be banned in public places?'*

With the question you will be given some 'stimulus' material to help you and approximately one week to do some research.

What should you do with the stimulus material?

Read it through carefully and identify any scientific words you do not understand. Look up the meaning of these words. The glossary in this book might be a starting point. Then go through with a highlighter pen and highlight those parts of the stimulus material you might want to use to answer the question.

What research should you do?

You should be looking for at least two or three sources of information. These could be from books, magazines, the Internet or CD-ROMs. You could also use surveys or experiments.

You will need to include with your report a list of sources that are detailed enough that somebody could check them. Some of your sources could look like this:

1 http://www.americanheart.org/presenter.jhtml?identifier=3016321

2 Heinemann, *Gateway Science OCR Biology for GCSE*, pages 51–152

3 *The Times*, 26th October 2005, page 4.

You can take this research material with you when you have to write your report about one week later. Do not print out vast amounts of irrelevant material from the Internet because you will not be able to find what you want when you write your report. Your teacher will probably collect in your research material to help them to assess your report but they will not actually mark it.

If you choose to do no research it does not stop you writing a report but you will get a lower mark.

Writing your report

Your will have to write your report in a lesson supervised by the teacher. It has to be your own work. There is no time limit but if you need longer than the lesson allowed, all work must be collected in and stored securely until the next time.

Your report should be between 400 and 800 words. As you write your report you need to refer to the information you have collected. It is all right to copy a short section from a source providing it is relevant and you have given credit for the source. For example:

The survey of 29,361 diners, conducted from May through mid-July, found that 96 percent are eating out the same amount or more often as a result of New York City's smoke-free restaurant law. Specifically, 23 percent said they were eating out more often because of the law. Seventy-three percent said they were eating out the same. Only 4 percent said they were eating out less often.

(http://www.americanheart.org/presenter.jhtml?identifier=3016321)

You should be critical of the sources. You will soon find that everything you read in newspapers or on the Internet is not necessarily true.

Make sure you answer the question. When you finish, read your report through carefully. Your teacher may give you details of the criteria they are using to mark your report.

Marking your report

Your teacher will mark your report against simple criteria and look for six skills, each marked out of a maximum of six. This makes a total of 36. Your teacher will explain these criteria to you.

This report is worth about 20% of the total marks. Hopefully, writing this report will make you more aware of science in our everyday lives.

Research study

The 'Research study' for GCSE Biology is similar to the 'Science in the news' study you may have already tried.

The 'Research study' for GCSE Biology involves looking at a scientific issue. You will concentrate on the work of scientists. They may be scientists working today or they may be scientists who worked in the past.

Your teacher will have a bank of tasks they can choose from. You only have to do one. If you do more than one the best mark will count.

Stimulus material

At the start of your 'Research study' you will be given some stimulus material. You should read this carefully. There are five questions that you will need to answer.

You will then be directed to do some research to extend the information you have. This research can be from books, CD-ROMs, the Internet etc. You should carefully reference where the information has come from. Your references can look like this:

1 http://www.enzymes.co.uk/

2 Heinemann, *Gateway Science OCR Biology for GCSE*, pages 75–77

3 Multimedia Science School 11–16 Enzymes

Answering questions

In a later lesson, under the supervision of your teacher, you will have to write a report answering the five questions you were given. You can take in any notes you made doing research. The questions must, however, be answered in the lesson. Your notes may be collected by your teacher but they will not be marked. You should not answer the questions in advance.

The questions you are asked are graded in difficulty. The first one or two questions are straightforward and use the information you were given at the start. They are intended for candidates who will get a grade E–G. The last one or two questions will be much harder and often there are alternative acceptable answers. These questions are intended for candidates who will get A*–B overall.

The report you write to answer these questions should be between 400 and 800 words. Many of the best reports that are produced are brief, but they answer the questions clearly without including unnecessary material.

Your teacher will mark your report. A sample of these will be checked later by a moderator from OCR to make sure the marking is fair. Your teacher will mark your report against four criteria on a scale of 0–6. The total mark is 24.

The table summarises these criteria.

Criteria	Advice to you
The evidence you have collected	You should try to collect evidence from at least two sources
How you have used the evidence to answer the questions	You should show how you have used the evidence to answer the questions
How the evidence helps you to understand scientific ideas	You should show how the evidence helps to explain scientific ideas, e.g. how ideas have developed over time how they are linked with social, economic and environmental issues
The quality of your report	You should be careful with your spelling, punctuation and grammar. Try to use correct scientific words

Your teacher may give you a set of 'student speak' criteria to help you.

Scientific ideas and how they develop

Throughout the work you are doing you should be aware of how scientific ideas develop over time. You will find examples in this book.

Where do you go from here?

Hopefully, having completed your course in GCSE Biology, you might want to study Biology further in the sixth form. It is important that you realise how scientific ideas have changed over time. It may well be that some of the ideas you have learned today may change as you study science in the future.

Biology is not a set of known facts that must be learned and passed on to future generations. Biology is a living subject that is likely to change as we find out more. Scientists have a responsibility to use this information for the good of everybody.

Good luck with your 'Research study'.

Data task

What is a 'Data task'?

During your GCSE Biology course you will probably do a task that involves analysing and evaluating some real data from an experiment. This is called a 'Data task'. You will also then do some planning of a further experiment.

Your teacher will have a number of 'Data tasks' they can give to you to do. You only need to do one. If you do more than one, your best mark will count.

An example of a 'Data task' is: *'How does temperature affect the working of an enzyme?'* This links with the science in B3a Molecules of Life.

Carrying out

You will be given some instructions to follow to do a simple experiment. You can do this individually or as part or a group. Alternatively, you can watch your teacher do a demonstration, or get some data from a computer simulation. You will need to collect some results and record these in a table.

There are no marks for collecting the results, but later on you are going to suggest improvements to the experiment. You cannot really do this unless you understand what was done in the first place.

Your results should be collected in by your teacher to keep them for the next lesson. If you didn't get any results, your teacher can give you a set. In the next lesson, your teacher will give you back your results and another sheet of questions.

You can write your answers in the spaces given on the sheet or you can write your answers on lined paper.

Writing up

This has to be done in a lesson supervised by your teacher. The separate additional sheet will usually tell you to:

1 Average your results.

2 Draw a graph to display your results. Remember your graph should fill at least half the grid. Make sure you choose a suitable scale for each axis and label each axis clearly.

3 Look for any pattern in these results.

4 Make some comments about the accuracy and reliability of your results. Look back at the table. Are the results the most accurate results you could get with the apparatus you have used?

For example, if you are using a hand-held stop watch it might show the time to the nearest one hundredth of a second, e.g. 10.17 s. But you would be better recording this as 10.2 s. There is a delay when you turn the stopwatch on and off and this makes a reading to better than 0.1 s wrong.

If you are using a burette to the nearest $1 \, cm^3$, you are not using it to the maximum accuracy. You should be able to read to the nearest $0.1 \, cm^3$. If you have three very similar results, e.g. 32.5, 32.4 and 32.5, this indicates that your results are reliable.

Also, if you look at the graph you have drawn, are all the points you have plotted either on or close to the line or curve? Again, this suggests reliability. Any points that are away from the graph are called anomalous results. You should be able to identify anomalous results. You should show these clearly on the graph or in your writing. Remember that if the results you collected are, for example, 32, 33 and 154, then 154 is an anomalous result and you should not include it. Instead you should ignore the 154 and average 32 and 33.

You may be given the opportunity to suggest what you could do to improve the experiment or get better results. This is called evaluation. Comments like 'take more readings' or 'do the experiment more carefully' are not worth credit, unless you qualify them.

5   At this stage, you should try to use some science to explain the pattern in the results you have found.

6   Finally, you will be asked to do some planning for a further experiment. This may be either to improve the experiment you have done or to extend the experiment to investigate another variable.

Marking your work

Your teacher will mark your 'Data task' against a set of criteria. He or she may give you a set of 'student speak' criteria. There are five things to be assessed by your teacher on a scale of 0–6. This makes the total for the 'Data task' a mark out of 30. This represents nearly 17% of the marks for the GCSE Biology award.

Criteria	Advice to you
Interpret the data	Can you draw a bar chart or a line graph to display your results?
Analyse the data	Can you see a pattern? This should be expressed as: As _____ increases, _____ increases. You might then be able to go further and explain the relationship.
Evaluate the data	Can you comment on the quality of the data and suggest any limitations of the method used?
Justify your conclusions	Can you link your conclusions with science and understanding?
Ideas for further work	Can you give a plan which is detailed enough for another person to follow it up?

Good luck with your 'Data task'!

Assessment of your practical skills

During your GCSE Biology course your teacher will have to make an overall assessment of your practical work. This is not based on any one practical activity, but is a general view of your practical work throughout the course.

There are two things your teacher will be asked to look for:

- How safely and accurately you carry out practical activities in science.

- How you collect data from an experiment, either individually or in a group with others.

Teachers are asked to use a scale of 0–6 and are given some help to do this.

They are told what is required for 2, 4 and 6 marks. They can give 1, 3 or 5 on their own judgements. If you have done no worthwhile practical work you may get 0.

The table summarises what is required for 2, 4 and 6 marks.

Number of marks	What is required?
2	You carry out practical work safely and accurately, but you need a lot of help doing the work.
4	You carry out practical work safely and accurately, but you need some help doing the work.
6	You carry out practical work safely and accurately and you do not need any help doing the work. You are aware of possible risks and take this into account.

This assessment is worth about 3.3% of the marks available for GCSE Biology.

Don't worry about asking for help thinking it might cause you to be marked down. The most important thing is you are able to complete the activity safely.

Enjoy the practical work in physics. If you go on to study Biology at AS and A2, the skills you have developed at GCSE will be very useful.

Good luck!

Glossary

acid rain rain with a pH below about 6 formed when pollutants such as sulfur dioxide and nitrogen oxides dissolve in it

acrosome a structure in the head of the sperm that contains enzymes to digest a pathway into the egg

active immunity immunity developed by the body to foreign invading organisms

active site a hole or groove on an enzyme molecule where the substrate enters

active transport the movement of substances across a cell membrane using energy from respiration, usually occurring against a concentration gradient

adapt change the characteristics of an organism, making it well suited to living in a particular environment

ADH (antidiuretic hormone) a hormone, secreted by the pituitary gland, which causes the kidney to reabsorb water when a person is dehydrated

adrenaline a hormone, secreted by the adrenal gland, which increases the heart and breathing rates when an organism is stressed

aerobic respiration respiration with oxygen

agglutinin a substance which causes red blood cells to clump together and clot

alcohol alcohol [ethanol], one of the products of the anaerobic respiration of sugar by yeast

algal bloom a sudden increase in the population of algae

alginate bead method of immobilising an enzyme using a droplet of jelly-like material obtained from seaweed

allele a genetic instruction received from one parent. Alleles from both parents form a gene

alveoli (singular **alveolus**) small air sacs found in the lungs

amino acids the building blocks of a protein. Amino acids contain nitrogen and join together to make polypeptides

amniocentesis a method of obtaining cells from samples of amniotic fluid [the liquid surrounding the fetus] and using them for genetic testing

anaerobic respiration respiration without oxygen

antagonistic muscles muscles with opposing actions eg the biceps and triceps in the upper arm. Muscles produce movement by contraction and work in pairs

antibiotic chemicals produced by fungi which kill bacteria or stop them from reproducing

antibody chemical produced by the body's immune system to destroy foreign invading organisms

antigen a thing that is foreign to the body

antiseptic chemicals which kill cells and prevent infection, can be used to clean wounds

aorta the main artery that carries blood from the heart, out to the body

artery blood vessel that carries blood away from the heart

artificially inseminate the placement of concentrated sperm in a female by non-natural means

asbestosis a lung disease caused by the inhalation of asbestos fibres

aseptic technique a procedure that is performed under sterile conditions

aspirin an analgeisc [painkiller] which also 'thins' the blood by acting as a mild anticoagulant. Often taken to prevent coronary thrombosis

assaying technique method for measuring the concentration of a substance eg the amount of urea in a sample of urine

asthma a respiratory disorder, often triggered by an allergen, which causes the airways to narrow thus making breathing difficult

atrium the left and right atria are the two upper muscular chambers of the heart; they pump blood into the ventricles

auxin a plant hormone that is produced in the growing points; it stimulates the growth of a shoot

average growth chart a medical chart comparing the growth rates of a large sample of people. It can be used to diagnose developmental problems in children

AVN atrio-ventricular node. This is a collection of nerves between the atria and ventricles of the heart. It triggers the contraction of the ventricles

axon the long extended part of a nerve cell

bacteria (singular **bacterium**) simple single-celled organisms which lack a true nucleus and which lack membrane-bound organelles such as chloroplasts and mitochondria

bases the four chemicals A, C, T and G that code for the instructions for life in DNA

battery farming keeping animals in controlled conditions indoors

benign not malignant. Will not grow like a cancer

biceps a muscle at the front of the upper arm. When the biceps contract the arm bends, an action known as flexion

bicuspid a valve between the left atrium and left ventricle in the heart

bile liquid that emulsifies fats, produced in the liver and stored in the gall bladder

binary fission division in which a single-celled organism divides into two identical daughter cells

binocular vision using both eyes to view the same object so that distance can be judged more accurately

binomial naming organisms with two names, one for genus and one for species

biodiversity range of different kinds of living organisms

biofuel fuels produced from natural sources that are renewable, for example, wood, alcohol from fermentation by yeast, and biogas from fermentation by bacteria

biogas a flammable mixture of gases formed by the anaerobic breakdown of dead organic matter

biological control the control of pests using living organisms

blood donor person who gives blood for the benefit of another

blood plasma the yellowish liquid which remains after blood has clotted

blood transfusion where blood from a donor is introduced into the circulation of a recipient, often after injury or after an operation

bypass surgery an operation in which a blood vessel, usually from the leg, is used to divert blood around a blockage in one of the coronary blood arteries

capillary tiny blood vessel that carries blood to the tissues of the body; a human being has thousands of miles of capillaries

capsule the part of a kidney tubule which surrounds the glomerulus [a knot of capillaries]. It collects the liquid produced by ultra-filtration

carbohydrase enzyme that digests carbohydrates

carbon dioxide a gas produced by respiration by both plants and animals and then used by plants for photosynthesis

carbonates often insoluble salt containing carbon and oxygen; calcium carbonate occurs naturally in limestone and chalk rocks, and also in coral reefs

cardiac cycle this describes one complete pumping operation of the heart. It includes the filling and emptying of the atria and ventricles

carnivore an animal which eats only meat

cartilage a white, shiny, slippery substance found on the moving surfaces of joints. It allows smooth, almost friction-free movement

cellulose substance made by plants that forms the structure of their cell walls

chemical digestion breaking food down using enzymes

chlorophyll a green pigment produced by plants that is used to trap light energy for the process of photosynthesis

cholesterol a fatty material mainly made in the body from saturated fat in a person's diet

chromosome structure composed of DNA and found in the nucleus of cells

cilia small hair-like structures on the surface of cells

cirrhosis disease where the liver becomes damaged

climate change changes in the climate such as global warming, brought about by the activities of humans

clone two or more organisms that are genetically identical

colour blindness not being able to see certain colours such as red and green

competition continual struggle that organisms have with each other for resources

complementary base pairing one type of base will always pair up with a particular other base in a DNA molecule

concentration gradient the variation in concentration of a substance in two different areas

contraception prevention of pregnancy

contractile vacuole an organelle which excretes surplus water from a cell

crenation shrinking of red blood cells due to their loss of water by osmosis

cross-breeding mating of two animals from different breeds

culture a population of micro-organisms growing on agar jelly or in a nutrient broth

cuticle a waxy layer mainly on the top surface of leaves that reduces water loss

cystic fibrosis a genetic disorder which results in the production of sticky mucus which blocks the airways and parts of the digestive system

Darwin the man who first put forward the theory of evolution

decay (plants and animals) breaking down of plant and animal matter

deficiency a lack of one or more minerals resulting in a lack of healthy growth

dehydration when an organism becomes short of water

denatured the change in shape of an enzyme molecule caused by high temperature or extremes of pH

dendrite cellular extension of nerve cells

denitrifying bacteria microbes that break down nitrates to nitrogen gas

depressant drug that depresses neural activity

detritivore an organism which feeds on dead or decaying organic matter eg leaf mould

detritus pieces of dead and decaying material

diabetes disease in which a person does not produce enough insulin to control the level of sugar in the blood

dialysis where blood is passed over a selectively permeable membrane which acts as a 'molecular sieve' and removes waste products

diastolic pressure blood pressure when the heart is relaxing

dicotyledonous plant that has two seed leaves

differentiation the process by which cells become specialised for different functions

diffusion a movement of particles from an area of high concentration to an area of low concentration

diploid when the chromosomes in a cell occur in pairs

distillation a method of separating alcohol from water by boiling and condensing the vapour. Used in the production of spirits such as whisky or rum

DNA the molecules that code for the instructions to make a new living organism

dominant an allele that always expresses itself

dormancy the state in seeds or buds where development or growth is occurring very slowly

double circulation a system that occurs in mammals where the blood passes through the heart twice on each circuit of the body

double helix the shape of a DNA molecule consisting of two chains twisted into a spiral

Down's syndrome a genetic condition resulting from a faulty cell division. This results in the fertilised egg containing 47 chromosomes instead of the usual 16

drug chemical that produces a change in the body

dwarfism a developmental disorder resulting in diminished growth and development. Often due to a lack of growth hormone from the pituitary gland

ECG an electrocardiograph is a graphical representation of the electrical signals produced by the heart

echocardiogram an echocardiogram is a moving image of the heart produced using ultrasound waves. The 'echo' is used to create the image

ecological niche part of a habitat

ecosystem a system of interacting organisms that live in a particular habitat

effector an organ such as a muscle that causes a response to a stimulus

egestion process by which material is ejected from an animal

emulsification the process of breaking down fat droplets into smaller ones

endangered organisms that are in danger of becoming extinct

enzyme organic catalyst that speeds up the rate of a reaction

eutrophication this occurs in nutrient-rich lakes and rivers when plants grow very rapidly and then run out of nutrients. As they die and rot, the water loses most of its dissolved oxygen causing the animals to die

evolution adaptation of organisms to changes in the environment through natural selection

exponential an increase that becomes more rapid with time

extinct organisms that no longer exist on the Earth

fat food storage molecules made from fatty acids and glycerol

fermentation another name for anaerobic respiration. A term used in the brewing industry to describe the making of alcohol from sugar using yeast

fermenter large tank in which millions of micro-organisms are grown in order to make useful products such as enzymes, vitamins, hormones or single-cell proteins

fertilise/fertilisation the fusion of male and female sex cells

fertility the ability to fertilise when male and female sex cells fuse together

fetal screening a number of methods used to investigate the development of an unborn baby. These include amniocentisis or ultrasound scans

fetus an embryo that has developed to the point where it contains all the necessary structures needed to grow into a new individual

flaccid the state of a plant cell that loses shape due to a drop in internal water pressure in the vacuole

flagellum a 'cork-screw shaped' structure which some bacteria use for swimming. It rotates like a propeller

food preservation keeping organic matter such as food in conditions that stop it decaying

fossil preserved remains or cast of a dead organism

fructose literally 'fruit sugar'. A simple sugar found in plants and a main ingredient in honey

FSH follicle stimulating hormone. A hormone released from the pituitary which stimulates the production of an egg

galactose one of the two sugars in lactose 'milk sugar' [lactose = glucose + galactose]

gamete a cell involved in reproduction, such as an ovum or a sperm

gene a section of DNA that codes for one specific instruction

genetic code sequence of bases that code for the instructions to make a new living organism

genetic engineering moving genes from one organism to an other

geotropism a growth response in plants either towards or away from gravity

germ theory of disease this is Louis Pasteur's explanation that many diseases and infections are caused by microscopic organisms such as bacteria and fungi

gestation period the period of time between fertilisation and birth

giantism a growth disorder resulting from the over-production of growth hormone by the pituitary gland

gill the organ which a fish uses to obtain oxygen and get rid of carbon dioxide

glomerulus one of the thousands of knots of capillaries in the kidneys where utra-filtration of blood takes place

glucose a type of sugar that is produced by photosynthesis and used as an energy source in respiration. Literally 'grape sugar'

growth hormone a hormone, produced by the pituitary, which stimulates growth

guard cells two cells that control the opening and closing of a stoma

haemoglobin a red protein containing an iron atom that can combine reversibly with oxygen

haemophilia a genetic disorder in which blood fails to clot normally. The abnormal allele is located on the X chromosome making it more common in males

hallucinogen a mind altering drug

haploid when each cell only has one copy of a chromosome from each pair

heart and lung machine a machine which takes over the function of the heart and lungs during major chest surgery

heart assist device diseased or damaged hearts may not pump efficiently. The damaged heart can be helped by pacemakers or mechanical pumps

heart transplant when a damaged or diseased heart is replaced by one from a healthy donor

heat stroke a condition caused when the body overheats

heparin a drug, originally extracted from leeches, which prevents blood from clotting

herbivore an organism which feeds on plants

heterozygous a gene that consists of two different alleles

hole in the heart a developmental defect where the hole between the atria of a fetal heart fails to close properly so oxygenated and deoxygenated blood mix

homeostasis maintaining a constant internal environment within the body

homozygous a gene that consists of two identical alleles

host a live organism that is fed upon by a parasite

hybrid a cross between two true breeding parents

hydroponics growing plants without soil usually in water

hypothermia a lowering of the body's temperature

immuno-suppressive drug a drug which prevents rejection by acting on the body's natural defence systems

inbreeding breeding between organisms that are closely related

incubation period the time between catching an infection and showing symptoms during which the micro-organisms are multiplying rapidly

indicator species a species that indicates the quality of the habitat by its presence there

infertility a problem involving the male, female or both which prevents them from having children

insulin hormone produced by the pancreas that lowers the level of glucose in the blood

intensive farming trying to produce as much food as possible from a certain area of land

intercostal muscles the muscles between the ribs which bring about breathing by changing the volume of the chest

invertase another name for sucrase

invertebrate animal without a backbone

iron lung a machine which produces breathing movements by changing the pressure on the outside of the chest

IVF (in-vitro fertilisation) a method of artificial fertilisation in which eggs and sperm are mixed in a glass dish outside the body. So called 'test-tube babies'

key a means of identifying different organisms

kidney the organs which remove and eliminate urea from the blood

kwashiorkor disease caused by a lack of protein in the diet

lactic acid chemical that causes muscle fatigue and is produced during anaerobic respiration

LH luteinising hormone is produced when an egg has been released. It results in the production of progesterone which is needed for pregnancy

ligament ligaments join bones to bones and hold joints in place

ligase an enzyme which builds new strands of DNA before cells divide

limiting factor a factor such as light, temperature or carbon dioxide where a lack of it limits the rate of photosynthesis

lipase enzyme that digests fat

liver a brownish coloured organ underneath the diaphragm. It helps regulate blood sugar by converting it to glycogen and releasing it when needed

long sight when the eye focuses light behind the retina

lower epidermis cells making up the underside of a leaf

lumen central hollow cavity that allows the passage of water in the xylem vessels

lysis the bursting of cells such as red blood cells that have taken up too much water by osmosis

malignant tumour that consists of rapidly dividing cells and is cancerous

meiosis cell division that occurs when gametes are produced; it reduces the number of chromosomes from 46 to 23

menstrual cycle the monthly cycle in which eggs are produced and released and the uterus prepared for pregnancy

methane a hydrocarbon (CH_4), a main constituent of biogas

microvilli (singular **microvillus**) microscopic projections found on the surface of the villi in the small intestine

minerals inorganic substances needed in small quantities by the human body and plants for good health

mitochondria (singular **mitochondrion**) microscopic organelles found in the cytoplasm of plant and animal cells; they are the site of many of the reactions of respiration

mitosis cell division that produces identical copies of cells

motor neurone a nerve cell that carries instructions away from the brain

mule sterile cross between a male donkey and a female horse

multi-cellular organisms comprising more than one cell, and having differentiated cells that perform specialised functions

mutation a change to the structure of a gene or DNA caused by such things as chemicals, X-rays or radiation

mutualism a relationship between two organisms of different species in which both organisms benefit

natural selection a process in which organisms that are most suited to the environment survive and produce more offspring

negative feedback a control mechanism where a rise in the level of one substance brings about a reduction in the level of another eg oestrogen and FSH

neurone a name for a nerve cell

nicotine an addictive chemical that is found in tobacco

nitrifying bacteria microbes that convert ammonium compounds to nitrates

nitrogen-fixing bacteria microbes that convert nitrogen gas into nitrates and other nitrogen compounds that plants can use

oestrogen a hormone produced by the ovary that causes the lining of the uterus to thicken

oil liquid fat

organic farming growing crops or raising animals without the use of chemical assistance

organism a living animal or plant

osmosis the movement of water from an area of high water concentration to an area of low water concentration across a partially permeable membrane

ossification the process by which cartilage is converted into bone

osteoporosis a process in which the mineral content of bone is gradually reduced. A loss in bone density causes them to break more easily

ovulation release of an ovum from the ovary

oxygen debt caused by anaerobic respiration and the production of lactic acid. After exercise the debt has to be repaid to break down the lactic acid

oxyhaemoglobin a chemical compound formed by the combination of haemoglobin and oxygen that takes place in the lungs

ozone a form of oxygen which can absorb ultraviolet light from the Sun preventing it reaching the Earth

pacemaker an electronic device which stimulates and controls the heart beat. Sometimes used to describe the SAN of the heart [see SAN]

pain killer drug that provides relief from pain

palisade mesophyll large rectangular cells in the leaf that are the main site of photosynthesis

parasite organism that lives on or in another living organism causing it harm

partially permeable a membrane that allows small molecules like water to diffuse through but blocks larger molecules

passive immunity short lasting immunity that is gained by injecting other peoples' antibodies

pasteurisation heating a fluid to above 63°C in order to kill harmful bacteria. Often done to beer or milk to improve its keeping qualities

pathogen disease-causing organism

performance enhancer drug that increases performance

permeable allowing substances to pass through

pesticide a chemical that will kill a pest on crops

phloem tissue that transports dissolved food substances (sugars) around a plant

photosynthesis the process by which plants convert water and carbon dioxide into oxygen and glucose using the energy from the Sun

phototropism a growth response in plants either towards or away from a light source

phytoplankton the plant component of a floating community of small organisms in a lake or in the sea

pituitary gland a small gland attached to the base of the brain which controls the release of hormones from many other glands in the body

placenta a structure produced by the embryo that grows into the wall of the uterus to absorb nutrients for the growing baby

plankton small floating organisms [animals and plants] which drift in the upper layers of large lakes and oceans and form the basis of many aquatic food chains

plant hormones chemical messengers that control how plants respond to stimuli – light, gravity and water; auxins are plant growth hormones

plaque a build-up of fatty deposits in blood vessels

plasma pale-yellow liquid that forms the fluid part of the blood

plasmid a ring of DNA found in many bacterial cells. Used in genetic engineering

plasmolysis shrinking of plant cells that have lost water leading to the cell membrane coming away from the cell wall

pollution the presence in the environment of substances that are harmful to living things

population a group of organisms of one species living together in a habitat

predator an animal that hunts and kills other animals for food

prey an animal that is hunted and killed for food by a predator

progesterone a hormone produced by the ovary after the ovum has been released. It maintains the wall of the uterus during pregnancy

protease enzyme for digesting protein

protein large polymer molecule made from a combination of amino acids

protozoan single-celled animals such as entamoeba. A member of the Kingdom Protoctista

pulmonary relating to the lungs, for example, pulmonary artery and pulmonary vein

pyramid of biomass a diagram showing the relative mass of organisms at each trophic level of a food chain

pyramid of numbers a diagram showing the relative number of organisms at each trophic level of a food chain

receptor an organ or cell that receives an external stimulus

recessive an allele that only expresses itself if the dominant allele is not present

recycling process by which materials are broken down, reprocessed and then reused, rather than being disposed of

red blood cell blood cell that contains the red pigment haemoglobin, which carries oxygen from the lungs around the body

reflex arc the pathway along which nerve impulses pass in a simple reflex action

rejection the process by which the immune system attacks foreign tissue such as a transplanted organ and causes it to die

relay a nerve cell that passes an impulse between two other nerve cells

residual air the volume of air remaining in the lungs and trachea [windpipe] when a person exhales as fully as they can

resources chemicals and materials that can be used for the benefit of humans

respiration a process that takes place in living cells converting glucose and oxygen into water and carbon dioxide with a release of energy

restriction enzyme an enzyme which cuts DNA into shorter lengths. It recognises certain sections of DNA and cuts only at these points

rooting powder a treatment containing plant growth hormones that is used to encourage cuttings to produce roots

SAN sino-atrial node, the heart's 'pacemaker'. This group of nerve cells controls and co-ordinates the beating of the heart

saprophytes bacteria and fungi that feed on dead organic material

saprophytic nutrition feeding on dead organic material by releasing enzymes and then taking up the soluble food

selective breeding a way of improving stock by selecting and breeding from those animals and plants that have the desired characteristics

selective reabsorption occurs in the kidney where materials needed by the body are removed from the filtrate produced by ultrafiltration in the glomerulus

selective weedkiller artificial plant hormones that kill some plants but not others

semilunar a half moon shaped valve at the beginning of each of the arteries leading from the heart

sensory neurone a nerve cell that carries information to the brain

sex hormone hormone that controls the secondary sexual characteristics

sheath the fatty coat that surrounds a nerve cell

short sight when the eye focuses light short of the retina

single circulatory system a circulatory system in which the blood must travel through the heart once for every complete circuit of the body eg in a fish

species a group of organisms that breed and produce fertile offspring

spirometer a machine which is used to record the movements of the chest and rate of use of oxygen during breathing

spongy mesophyll a layer of cells in the leaf that have large airspaces between them to allow gases to diffuse

starch a carbohydrate food storage substance produced by plants

stem cells cells that have not differentiated and can still produce different types of cells or tissues

sterilised heated until all forms of life have been destroyed. Other methods of sterilisation include the use of radiation or chemicals [antiseptics or disinfectants]

stimulant drug that stimulates neural activity

stomata (singular **stoma**) small pores on the underside of a leaf that regulate the release of water, and allow the release of oxygen and the absorption of carbon dioxide

substrate a substance acted upon by an enzyme

sucrase the enzyme which digests sucrose to form glucose and fructose

sucrose a 'double sugar' consisting of glucose + fructose

surrogate a person or animal that acts as a substitute mother for another by having an embryo implanted

sustainable development development using the Earth's resources at a rate at which they can be replaced

synapse a minute gap between two neurones

synapse an exceedingly small gap between two neurons

synovial joint a moveable joint eg a hinge joint or ball and socket joint which is lubricated by a fluid

systolic pressure blood pressure when the heart is contracting

tidal air the volume of air passing in and out of the lungs during breathing. The amount varies with activity

tissue culture using small pieces of organisms to grow genetically identical organisms

toxin poison produced by microorganisms

transgenic an organism which contains DNA from different species eg bacteria containing genes to manufacture human insulin

translocation the movement of sugars and other food materials through the phloem in plants

transmitter a chemical that transmits a nerve impulse across a synapse

transmitter substance a chemical that is released and diffuses across the gap between two neurones in a synapse

transpiration the loss of water from plant leaves

triceps a muscle at the back of the upper arm. When the triceps contract the arm straightens, an action known as extension

tricuspid a valve between the right atrium and right ventricle in the heart

trophic level a feeding level in a food chain

tubule part of the kidney where substances such as water salts and glucose are reabsorbed back into the bloodstream

tumour a group of unspecialised cells that may be malignant or benign

turgid the state of a plant cell when it is held in shape by water pressure in the vacuole causing it to press against the cell wall

upper epidermis the top layer of cells in a leaf; it does not contain any chloroplasts

urea a component of fertiliser and animal feed

urine the liquid waste, consisting of water and urea, which is excreted by the kidneys

vascular bundle strands of vascular tissue made up of xylem and phloem

vasoconstriction a decrease in the diameter of blood capillaries

vasodilation an increase in the diameter of blood capillaries

vector organism that transmits and carries a disease

vein (animal) a blood vessel that returns blood to the heart

vein (plant) the part of a leaf that carries water to the leaf cells

vena cava the large vein that returns blood from the body to the heart

ventricle the lower two muscular chambers of the heart; the left ventricle pumps blood around the body; the right ventricle pumps blood to the lungs

vertebrate animal that has a backbone

villi (singular **villus**) small finger-like projections in the small intestine that increase the surface area for absorption

vital capacity the maximum amount of air which can be taken into the lungs and breathed out in one large breath

warfarin an anticoagulant, a drug which prevents blood from clotting, which is taken by people at risk of heart attacks or strokes

white blood cell found in the blood and forms part of the body's defence mechanism; they produce antibodies and engulf bacteria

wilting what happens to a plant that does not take in enough water by osmosis

xylem tissue that transports water and dissolved minerals around a plant

yeast a single-celled fungus used in the brewing and bread making industries

zygote the single cell produced when two gametes join

Index

Revision Guides

Beat the rest - exam success with Heinemann

Ideal for homework and revision exercises, these differentiated **Revision Guides** contain everything needed for exam success.

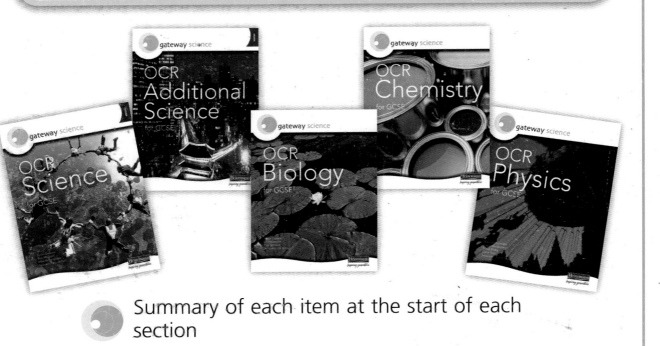

- Summary of each item at the start of each section

- Personalised learning activities enable students to review what they have learnt

- Advise from examiners on common pitfalls and how to avoid them

Please quote S 603 SCI A when ordering

t 01865 888068 **f** 01865 314029 **e** orders@heinemann.co.uk **w** www.heinemann.co.uk

Heinemann

Inspiring generations

L554